T0348442

Fourier Transform Spectrometry

Fourier Transform Spectrometry

Sumner P. Davis

Berkeley, California, U.S.A.

Mark C. Abrams

Fort Wayne, Indiana, U.S.A.

James W. Brault

Louisville, Colorado, U.S.A

ACADEMIC PRESS

An Imprint of Elsevier

San Diego San Francisco New York Boston

London Sydney Tokyo

This book is printed on acid-free paper. ♾

ACADEMIC PRESS

An Imprint of Elsevier

525 B Street, Suite 1900, San Diego, California 92101-4495, USA

http://www.academicpress.com

Academic Press

Harcourt Place, 32 Jamestown Road, London NW1 7BY, UK

http://www.academicpress.com

Library of Congress Catalog Card Number: 2001089161

ISBN-13: 978-0-12-042510-5
ISBN-10: 0-12-042510-6

Transferred to Digital Printing 2008

05 06 IP 9 8 7 6 5 4 3 2

This book is dedicated to our wives,
who put up with "The Book"
for more than a decade.
Without their patience, support,
and gentle encouragement
this book would have never progressed
beyond a dream.

TABLE OF CONTENTS

The cover image: A high-resolution digital spectrum of the Sun between 4000 and 7000 angstroms displayed in the form of an echelle spectrogram. Nigel Sharp (National Optical Astronomy Observatory/Kitt Peak National Observatory) colorized the digital solar atlas observed with the 1-m Fourier Transform Spectrometer at the McMath–Pierce Solar Facility on Kitt Peak by Robert L. Kurucz, Ingemar Furenlid, James Brault, and Larry Testerman (*National Solar Observatory Atlas No. 1, June 1984*) and sliced it into orders similar to those produced by an echelle. This image is available electronically on the NOAO Image Gallery (*http://www.noao.edu/image_gallery*).

PREFACE

Fourier transform spectroscopy has evolved over several decades into an analytic spectroscopic method with applications throughout the physical, chemical, and biological sciences. As the instruments have become automated and computerized the user has been able to focus on the experiment and not on the operation of the instrument. However, in many applications where source conditions are not ideal or the desired signal is weak, the success of an experiment can depend critically on an understanding of the instrument and the data-processing algorithms that extract the spectrum from the interferogram. *Fourier Transform Spectrometry* provides the essential background in Fourier analysis, systematically develops the fundamental concepts governing the design and operation of Fourier transform spectrometers, and illustrates each concept pictorially. The methods of transforming the interferogram and phase correcting the resulting spectrum are presented, and are focused on understanding the capabilities and limitations of the algorithms. Techniques of computerized spectrum analysis are discussed with the intention of allowing individual spectroscopists to understand the numerical processing algorithms without becoming computer programmers. Methods for determining the accuracy of numerical algorithms are discussed and compared pictorially and quantitatively. Algorithms for line finding, fitting spectra to voigt profiles, filtering, Fourier transforming, and spectrum synthesis form a basis of spectrum analysis tools from which complex signal-processing procedures can be constructed.

This book should be of immediate use to those who use Fourier transform spectrometers in their research or are considering their use, especially in astronomy, atmospheric physics and chemistry, and high-resolution laboratory spectroscopy. We give the mathematical and physical background for understanding the operation of an ideal interferometer, illustrate these ideas with examples of interferograms that are obtained with ideal and nonideal interferometers, and show how the maximum amount of information can be extracted from the interferograms. Next, we show how practical considerations of sampling and noise affect the spectrum.

This book evolved out of 20 years of conversations about the methods and practice of Fourier transform spectroscopy. Brault is the author of several papers incorporated into this text, as well as author of a seminal set of lecture notes on the subject. He is the prime mover in establishing the Fourier transform spectrometer (FTS) as the instrument of choice for high-resolution atomic and molecular spectroscopy. The content of this book is taken mainly from his work in optics and instrumentation over a period of many years. Some of the text was initially written up as a part of Abrams's doctoral dissertation in order to clarify and quantify many rules of thumb that were developed in the field by Brault and others. Davis and his graduate students were early users of the Kitt Peak and Los Alamos instruments and have continually pushed for ever-greater simplicity, accuracy, and flexibility of the data-taking and data-processing procedures.

Sumner P. Davis is a Professor of Physics at the University of California at Berkeley. His research focuses on laboratory spectroscopy of diatomic molecules of astrophysical interest. Since the late 1950s, Davis and his graduate students pushed the limits of high-resolution molecular spectroscopy — initially with 1- to 12-m grating spectrometers and echelle gratings, then with high-resolution Fabry–Perot interferometers (crossed with high-resolution gratings), and since 1976 with Fourier transform spectrometers. Most recently, he has returned to echelle grating spectrometers — bringing full circle a five decade adventure in spectroscopy.

Mark C. Abrams is Manager of Advanced Programs for ITT Industries Aerospace /Communications Division in Fort Wayne, Indiana. He was a staff member at the Jet Propulsion Laboratory, California Institute of Technology, where he was the instrument scientist for the Atmospheric Trace Molecule Spectroscopy (ATMOS) experiment (a Space Shuttle-borne Fourier transform spectrometer used for Earth remote sensing). His research focuses on remote sensing from space and instrument design.

James W. Brault is a physicist and was a staff scientist of the National Solar Observatory, Kitt Peak, with an appointment at the University of Colorado in Boulder. He is the designer of the one-meter instrument at the Observatory, and a codesigner of the 2.5-meter spectrometer formerly at the Los Alamos National Laboratory and now at the National Institute for Standards and Technology (NIST). He has also pioneered numerical methods for transforming and reducing Fourier transform spectra. His other areas of research are atomic and molecular spectroscopy.

We owe much to our contemporaries and predecessors in the field. We gratefully acknowledge several kind colleagues and reviewers, Professor Luc Delbouille, Dr. Ginette Roland, Professor Anne P. Thorne, and Dr. Brenda Winnewisser, who gave their time generously and shepherded the book through many necessary changes. The inevitable mistakes are ours (with apologies, but enjoy them), and we hope that these pages will inspire a new generation of researchers to push beyond the current state-of-the-art and take the community forward with enthusiasm.

S. P. D., M. C. A., and J. W. B. — Spring 2001

1

INTRODUCTION

Electromagnetic radiation in the classical picture is a traveling wave of orthogonal electric and magnetic fields whose amplitudes vary in time. The propagation of the wave is described by the wave equation, which is derivable from Maxwell's equations. We will consider only the electric field, since the magnetic field amplitude and phase are linearly related to the electric field, and the observable effects on our detectors are largely due to the electric rather than the magnetic fields.

As we shall see, this wave carries information about the source that generates it. We can make a model of the source when we know the frequencies and their amplitudes and phases that make up the time-varying wave. Think of how a prism separates light into its constituent colors and how we can use it as a tool for examining the spectra of different light sources. It disperses the light by changing its direction of propagation proportionally to the frequency. Sunlight is spread into an almost uniformly intense visible spectrum from violet to red, while light from a high-pressure mercury lamp is concentrated in the violet and blue. A diffraction grating also disperses the light, but into several spectra rather than a single one, and through angles proportional to the wavelength rather than the frequency. Because we know how to describe the action of a dispersing element on an electromagnetic wave, we like to think that there is an underlying physical principle and mathematical process that describes the resolution of a time-varying wave into its constituent frequencies or colors, independent of what form the dispersing instrument takes. The process is called *Fourier decomposition of the waves.* The amplitude *vs.* time function is transformed into an amplitude *vs.* frequency spectrum. The decomposition technique is not restricted to electromagnetic radiation, but applies to all waves, including sound and water waves. Often the term *Fourier transformation* is used as a general term encompassing both Fourier synthesis and Fourier decomposition.

Let's pick an example to show what this transformation does. We will add three sinusoids of different frequencies and then decompose the combination into its three components. Strike a tuning fork lightly and we get a sinusoidal wave, in this case sound in air, say, at a frequency of 300 Hz. An amplitude *vs.* time plot is shown in Fig. 1.1. Strike a second fork and get a frequency of, say, 600 Hz, as shown in Fig. 1.1. A third fork, with a frequency of 1200 Hz, as shown in Fig. 1.1. Now strike them simultaneously and observe the beat frequency (Fig. 1.1).

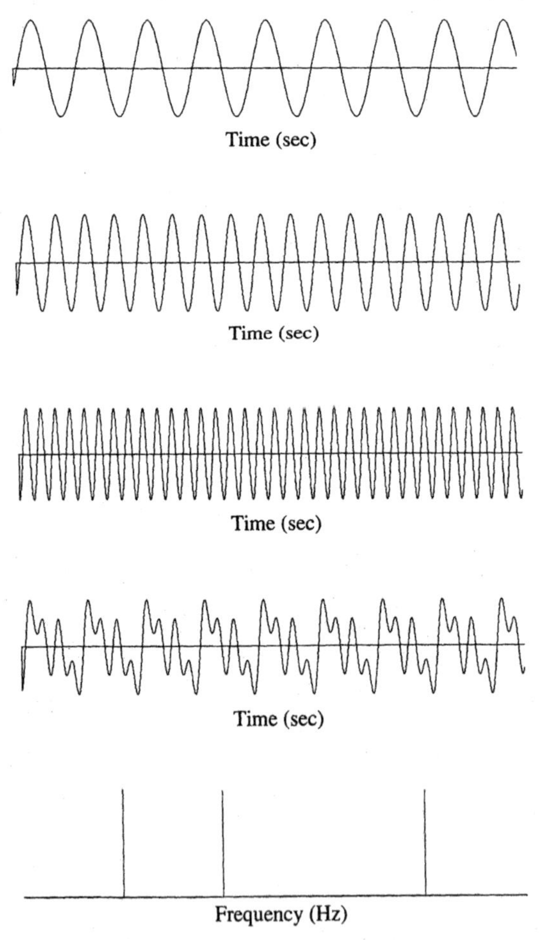

Fig. 1.1 Sinusoidal waves of frequency 300 Hz, 600 Hz, and 1200 Hz. All three waves added together. Fourier decomposition of the combined wave, showing the presence of three frequencies.

The complex wave looks very different from any of the three sinusoids. We have synthesized it from individual sinusoidal waves. If we see only the complex wave itself, we might not realize that it is three distinct frequencies combined. But when we execute a Fourier transformation, we get back a plot of amplitude *vs.* frequency as show in Fig. 1.1. This plot tells us that we have three sinusoids at 300, 600, and 1200 Hz. The transform does not plot out three sine waves for us, but only specifies the relative amplitudes of the frequency components illustrated in Figs. 1.1. When we listen to the sound given off by the three tuning forks struck simultaneously, we can actually hear the three frequencies making up the wave. The ear performs a Fourier transform.

In optics, there are many such devices, so many that we do not even think of them as instruments performing Fourier transformations on radiation. For examples, a thin film of oil on the ocean splits sunlight into its constituent colors, and we see these colors floating on the surface. A diffraction grating disperses light into different directions depending on the frequencies present. The amplitudes and frequencies tell us about processes in the source from which the radiation comes. In the cases we are interested in here, the sources are solids, liquids, or gases and the radiation emitters are atoms or molecules.

The observation and measurement of radiation emitted from or absorbed by atoms and molecules provides information about their identity and structure and their physical environment. A spectrum may contain discrete lines and bands and perhaps a continuum. Each feature can be identified and associated with a particular atom or molecule and its environment. For examples, spectrochemical analysis uses spectroscopic features to determine the chemical composition of the source of the radiation. Spectroscopists use the frequency distribution to measure the energy level structure within an atom or molecule. This structure dictates the unique interaction between radiation and emitting or absorbing matter. For a gas, measurements of spectral features can be interpreted in terms of abundances, temperatures, pressures, velocities, and radiative transfer in the material.

Spectrometry, as we are going to discuss it, is the detection and measurement of radiation and its analysis in terms of frequency and energy distribution, called the *spectrum*. The measurement of the spectrum involves the determination of spectral line positions, intensities, line shapes, and areas. Fourier transform spectrometry is a method of obtaining high spectral resolution and accurate photometry so that the measured intensity is an accurate representation of the radiation. We will use the abbreviation *FTS* to mean either Fourier transform spectrometry or Fourier transform spectrometer, as the occasion requires. What we measure ideally should represent the frequency spectrum of the radiation unmodified by any optical instrument or computational artifacts. In practice, accurate spectrometry is a result of understanding how the observed spectrum is modified by the observational process. To find the frequencies present in the radiation we could in principle record the intensity of the radiation as a function of time and take a Fourier transform to obtain the frequency spectrum. In practice, we cannot do this because we have no detectors and associated electronic circuitry that respond to optical frequencies. Instead, we use a scanning Michelson interferometer to transform the incoming radiation into amplitude-modulated radiation whose modulation frequencies are the scaled-down optical frequencies in the original signal and typically fall in the range of 0 to 50 KHz. We detect these modulation frequencies and record them as the interferogram. Our task is then to take the transform of this interferogram to recover the original frequency spectrum of the radiation. interferometer

Why go through these complicated transformations when there are optical instruments such as the diffraction grating spectrometer that transform incoming radiation directly into its individual spectral components? The FTS is the most accurate general-purpose passive spectrometer available. Even if not always the simplest or most convenient, it can be used to provide working standards for testing and calibrating more rapid but less precise techniques. In addition to its inherent precision, because it is an interferometer with a large path difference and a well-known instrumental function, the FTS is noted for high optical efficiency, no diffraction losses to higher-order spectra, high throughput, simultaneous observation of all frequencies/wavelengths, precise photometry, easily adjustable free spectral range, wide spectral coverage, and two-dimensional stigmatic imaging. Where spectral line profiles are important, the FTS offers the most thoroughly understood and simply characterized apparatus function of any passive spectrometer, because all radiation falls on a single detector and instrumental distortions are often accurately calculable and correctable.

Far more than most instruments, the Fourier transform spectrometer is a useful and practical realization in metal and glass of a simple and elegant mathematical idea, in this case Fourier's theorem. It is true that all of our instruments derive their usefulness from physical phenomena that can be described by equations, but their basis is usually well hidden by the time the instrument is in the hands of an experimenter. The complex processes involved are normally summarized in some kind of working equation, for example, the grating equation, which maps the desired variable, wavelength, onto the directly observed variable, angle or position at which the intensity is observed.

In Fourier spectroscopy, the observations are made in a conjugate space, and only produce recognizable results after a mathematical transformation. We observe on the other side of the equation, so to speak. Now, this transformation poses no difficulties for the *instrument*. The needed measurements are clearly defined and readily obtained, and the computer programs for making the transformation are simple and efficient. The real practical difficulty with this technique is that all of the instincts and experience of the *user* have come on the one side of the equation, so we must either operate blindly, and hence often foolishly, or find some way to carry our experience across the transform sign. The discussion that follows is aimed at developing some modes of thought and mathematical tools to make the latter path easier to follow. The treatment is brief for topics that are already well covered in articles or texts, and more complete for areas that are less well understood in practice.

Several extensive bibliographies are collected at the end of the book. The first is a chapter-by-chapter bibliography in which the references are presented by topic within each chapter. The next is a chronological bibliography; the third is an applications bibliography, covering laboratory work and applications to remote sensing. The last is an author bibliography.

1.1 Spectra and Spectroscopic Measurement

The decomposition of electromagnetic radiation into a spectrum separates the radiation into waves with various frequencies and corresponding wavelengths: $\nu\lambda = c$, where ν is the frequency in hertz (Hz) and λ is the wavelength, with c the speed of light. For optical spectroscopy in the infrared (IR), visible, and ultraviolet (uv) regions of the spectrum, it is customary to work with the wavelength λ measured in a length unit of nanometers (10^{-9} m), abbreviated nm, or in a length unit of micrometers (10^{-6} m), abbreviated μm, and with the wavenumber $\sigma = \nu/c = 1/\lambda$ measured in waves/centimeter in vacuum, abbreviated cm^{-1} and termed *reciprocal centimeters*. However, the latest recommendation by an international commission is to use σ for the wavenumber in a medium, and $\tilde{\nu}$ for vacuum wavenumber or without the tilde "when there is no chance of confusion with frequency." We will consistently use the first definition of σ in this book and note that it is sometimes useful to consider σ as a spatial frequency.

The spectrum is partitioned into broad regions as illustrated in Fig. 1.2. Which of the preceding terms is used in any given case depends on the spectral region, the observational techniques, and the particular phenomenon being investigated.

Fourier transform spectrometry covers a significant portion of this spectrum, from 5 cm^{-1} in the far infrared to 75 000 cm^{-1} in the vacuum ultraviolet. FTS instruments are distinguished by an ability to cover broad spectral ranges with high resolution. A single scan can collect spectral data over ranges as large as 10 000–30 000 cm^{-1} simultaneously. While other spectrometers can cover similar spectral ranges, the range of a single set of measurements is typically limited to 10% or less of the potential range of the instrument class.

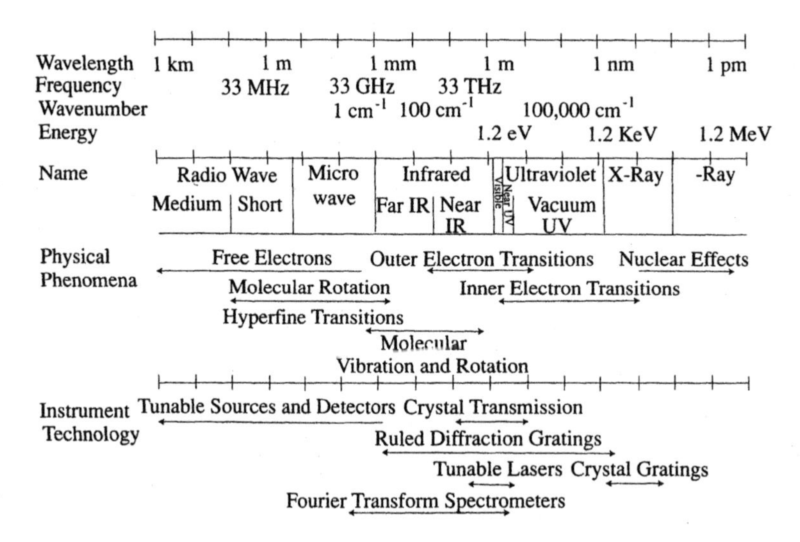

Fig. 1.2 Frequency, wavelength, and wavenumber ranges for the measurement of electromagnetic radiation.

Fundamentally, this book is about methods developed for measuring the best possible spectra of laboratory, terrestrial, and astrophysical objects. Each of us migrated to FTS as a means of obtaining the best spectra, rather than developing a technique and seeking applications for that technology. It is spectra that have captured our attention, and consequently various spectra will appear throughout this book.

These sample spectra are intended to be representative rather than comprehensive, and they reflect our personal biases. Because they are spectra we obtained in our research, they also have the convenience of proximity and accessibility. With the advent of the Internet, many of the historic libraries of spectra that were locked away in archives are now accessible to the general public. The electronic archive at the National Solar Observatory (NSO), Kitt Peak, in which the McMath–Pierce FTS spectra are archived, is a unique record of the atomic, molecular, solar, and stellar spectra that have provided a major contribution to modern spectroscopy in the last 25 years (perhaps 25,000 spectra). Similarly, the Atmospheric Trace Molecule Spectroscopy (ATMOS) experiment (a Space Shuttle-based FTS), used for atmospheric profiling and solar observations from space, provided some 80,000 high-resolution spectra in the 2- to 16-micrometer spectral region. Indeed, the two laboratories supported each other, with the HITRAN (high-resolution transmission) molecular spectroscopy database summarizing the results of years of measurements at NSO that provided the theoretical basis for modeling the Earth's atmosphere.

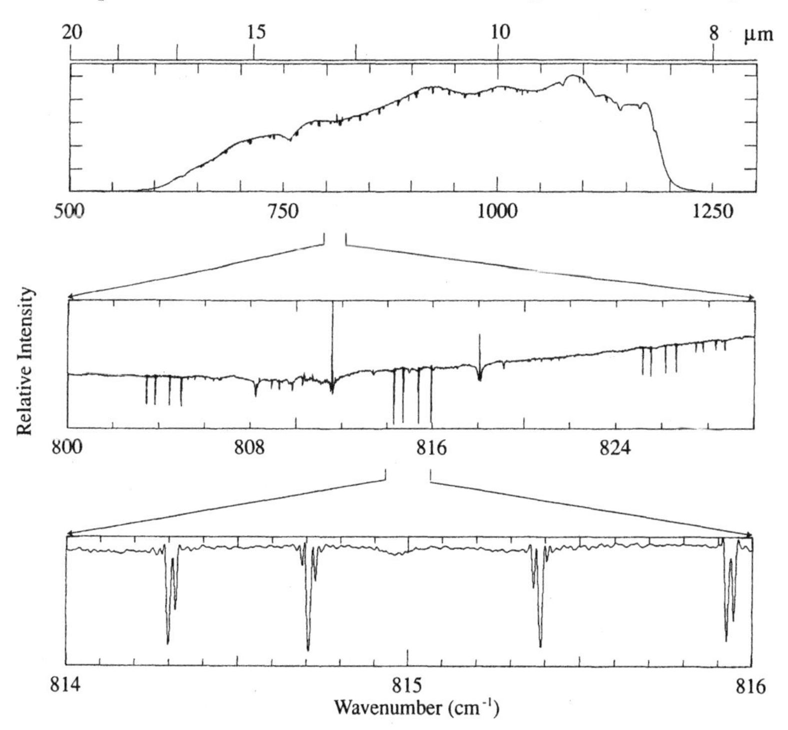

Fig. 1.3 A solar absorption spectrum as observed from the Space Shuttle. The Atmospheric Trace Molecule Spectroscopy (ATMOS) experiment obtained high-resolution solar absorption spectra between 625 and 5000 cm^{-1} (2–16 micrometers) with signal-to-noise ratios of 300–400, as illustrated in the traces.

High-resolution infrared solar spectra have remained a major research topic throughout the past 30 years. Of the 16 000 detected solar features, the majority are lines of vibration-rotation bands of the diatomic molecular constituents of the photosphere: CO, CH, OH, and NH. The $\Delta v = 1$ and $\Delta v = 2$ bands of $^{12}C^{16}O$ and its isotopic variants dominate the spectrum, about 55% of the lines. About 11% of the lines are due to atomic transitions in neutral atoms of Fe, Si, Mg, C, Ca, and Al. But even after 15 years of examination, 24%, or 3700 lines, remain unidentified.

In the long-wavelength infrared spectral region (10–15 micrometers) the solar spectrum shown in Fig. 1.3 is dominated by the rotational bands of the hydroxyl radical and the Mg emission lines. Since their discovery, the Mg emission lines, which are Zeeman-split by the magnetic fields present on the surface of the Sun, have been extensively studied for magnetic field mapping.

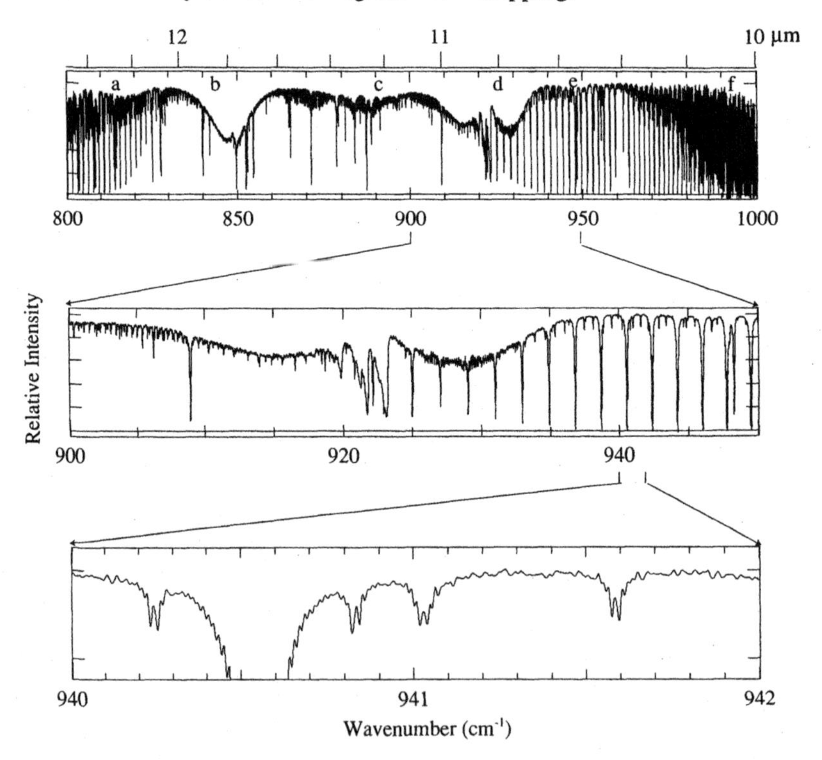

Fig. 1.4 Solar absorption spectra of the Earth's atmosphere illustrating the absorption features of various trace gases: (a) CO_2, (b) CFC-11, (c) HNO_3, (d) CFC-12, (e) CO_2, (f) O_3.

Solar absorption spectra of the Earth's limb during sunrise and sunset using the Sun as a source, provide long-path observations from which the composition and state of the Earth's atmosphere can be measured. They are shown in Fig. 1.4. The low-dispersion spectrum illustrates absorption features due to carbon dioxide, ozone, and nitric acid, which are essential components of the atmosphere, with concentrations of parts per thousand, parts per million, and parts per billion, respectively. In addition, the absorption features of the anthropogenic refrigerants CFC-11 (CCl_3F) and CFC-12 (CCl_2F_2) are clearly visible in this tropospheric spectrum at 10-km tangent altitude. The noise level is approximately the vertical excursions of the trace between spectral lines in the high-dispersion lower panel. The periodic ripple apparent in this panel is molecular absorption of HNO_3. Theoretical calculations do a remarkable job of simulating the infrared spectrum, as evidenced by the work summarized in the HITRAN database.

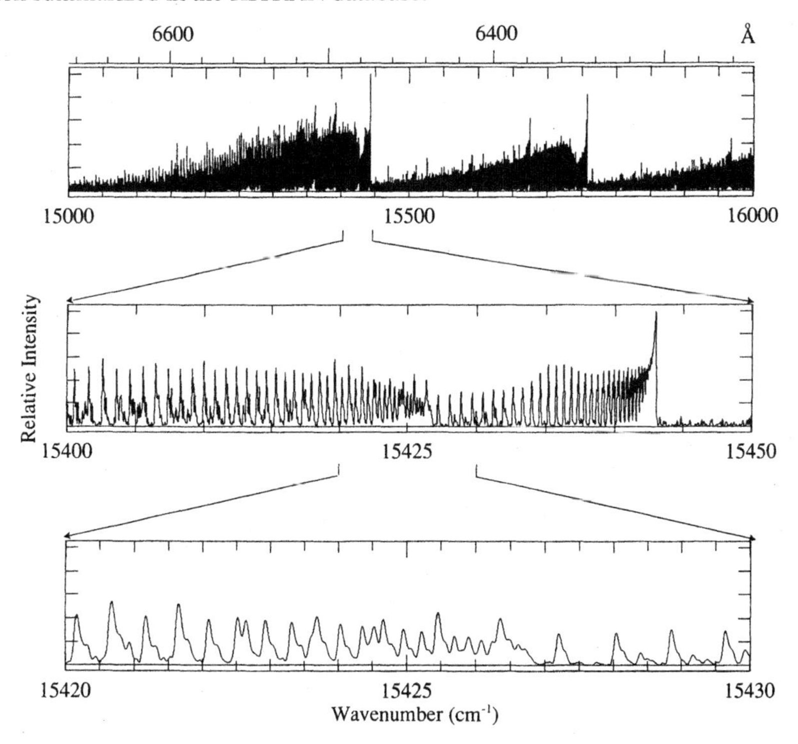

Fig. 1.5 Emission spectrum of ZrO in a furnace. The traces shown are all plotted from the same observational broadband run and illustrate the fact that a broad spectral region can be covered at high resolution in a single scan.

Simulation of stellar environments with a high-temperature ($T > 2000$ K) furnace is a method of observing molecular emission and absorption under controlled conditions. The visible spectrum of zirconium oxide shown in Fig. 1.5 is observed in the spectra of cool carbon stars. Laboratory measurements of line positions, amplitudes, and equivalent widths permit accurate simulation of spectra in these stars and the determination of the structure, composition, and evolution of the stars. The FTS enables accurate and consistent broadband high-resolution spectrometry covering large intensity ranges. With a properly calibrated FTS and spectral measurement software, such measurements are routine and performed on a regular basis. The catalog of atomic and molecular parameters measured with the NSO FTS is remarkable in regard to both its sheer volume and the duration of this instrument as a standard for laboratory measurements over its 25 years of operation.

1.2 The Classical Michelson Interferometer

We start with a Michelson interferometer, as shown in Fig. 1.6. Light from a source at the object plane is collimated and then divided at a beam splitter into two beams of equal amplitude.

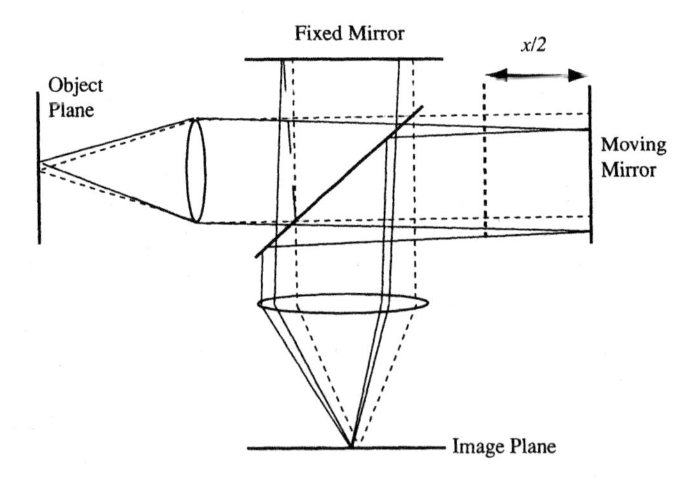

Fig. 1.6 Sketch of Michelson interferometer with an extended source.

These beams are reflected back on themselves by two separate mirrors, one fixed and the other movable. Each single beam strikes the beam splitter again, where they are recombined and split again. Consider only the recombined beam that is directed to the image plane. The two components in the recombined beam interfere with each other and form a spot whose intensity depends upon the different paths traversed by the two beams before recombination. As one mirror moves, the path length of one beam changes and the spot on the screen in the image plane becomes brighter and dimmer successively, in synchronization with the mirror position. Circular fringes rather than spots are formed when the source is an extended one.

For an input beam of monochromatic light of wavenumber σ_o and intensity $B(\sigma_o)$, the intensity of the interferogram as a function of the optical path difference x between the two beams is given by the familiar two-beam interference relation for the intensity,

$$I_o(x) = B(\sigma_o)[1 + \cos(2\pi\sigma_o x)] \tag{1.1}$$

where as noted earlier the wavenumber σ is defined by $\sigma = 1/\lambda = \nu/c$, measured in reciprocal centimeters. When x is changed by scanning one of the mirrors, the interferogram is a cosine of wavenumber, or spatial frequency, σ_o.

When the source contains more than one frequency, the detector sees a superposition of such cosines,

$$I_o(x) = \int_o^\infty B(\sigma)[1 + \cos(2\pi\sigma x)]\, d\sigma. \tag{1.2}$$

We can subtract the mean value of the interferogram and form an expression for the intensity as a function of x:

$$I(x) = I_o(x) - \overline{I(x)} = \int_o^\infty B(\sigma)\cos(2\pi\sigma x)\, d\sigma. \tag{1.3}$$

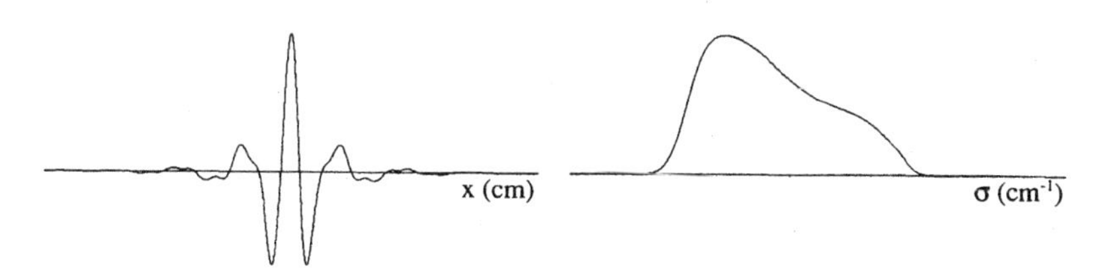

Fig. 1.7 Symmetric interferogram and the spectrum derived from its Fourier transform.

The right-hand side contains all the spectral information in the light and is the cosine Fourier transform of the source distribution $B(\sigma)$. The distribution can therefore be recovered by the inverse Fourier transform,

$$B(\sigma) = \int_o^\infty I(x)\cos(2\pi\sigma x)\, dx. \tag{1.4}$$

A schematic representation of an interferogram and the resulting spectrum are shown in Fig. 1.7.

1.3 Precision, Accuracy, and Dynamic Range

How precise are the measurements? Often we mean how many decimals in the frequencies are significant, but we can also apply the question to intensities, line shapes, and widths. In most current applications it is not enough to know only the peak position and the peak intensity. Line shapes and widths are important, and they influence the position and intensity. These parameters are accurately determined only when the lines are fitted to an appropriate profile, something easy to determine for the FTS. Fitting is also required because the point of maximum intensity almost never occurs at a sampling point. Our fitting codes assume a voigtian profile, and also account for the instrumental profile. With proper sampling and fitting we can measure a single unblended line position to a precision of the line width divided by the S/N ratio or even better. For example, a line at 2000 cm^{-1} with a width of 0.015 cm^{-1} and a S/N ratio of 100 can be measured to 0.00015 cm^{-1}, approximately 1.3 parts in 10 million.

How accurate are the measurements? When we assign a wavenumber value, how good is it on an absolute scale? Two considerations are important, the precision just discussed and the accuracy of our standards. At the present time our standards are provided by N_2, O_2, CO, Ar, and a few other spectrum lines, so these set the ultimate accuracy. Not every source operates with the same internal conditions nor do they radiate exactly the same frequencies, so source variations always affect the wavenumber accuracy. Current practice yields results accurate to one part in 10^7, while state of the art is an order of magnitude better. All this may change in the next few years because standards are continually improving and being extended to very large and very small wavenumbers.

It is in the realm of measuring line intensities, both absolute and relative, that the FTS is without a serious competitor. Atomic spectrum lines vary from line to line by orders of magnitude in intensity, and it is always the weak lines that verify the correctness of an energy level analysis. The ratio of the strongest lines to the weakest detectable ones is called the *dynamic range of detection.*

Lines from typical sources have dynamic ranges of 1000 or less, but for low-noise-emission sources, dynamic ranges approaching 1 million are occasionally encountered. Observing a dynamic range of 1000 is routine for FTS measurements, and with care a dynamic range of 30,000 can be achieved. Reaching this limit with other spectrometric techniques is challenging at best.

1.4 Units

The diversity of units in a given field is related to the longevity of the discipline, and optics and spectroscopy span nearly the entire period between Newton's discovery of the dispersive properties of prisms in 1672 to the present day. Consequently, there are more units than necessary.

Spectral plots are two-dimensional representations of energy as a function of wavelength or frequency. The wavelength and wavenumber scales and their units have been discussed in Section 1.1. Recall that the appropriate scale for Fourier transform spectrometry is wavenumbers measured in *reciprocal centimeters* or *inverse centimeters*, written as cm^{-1}. Two equivalents are $1\ cm^{-1} = 2.9979 \times 10^{10}$ Hz $\sim$ 30 GHz.

The intensity scale (y-axis) can be even more confusing. Once upon a time the scale was quantified in terms of *spectral radiance*, which was measured in units of erg sec^{-1} cm^{-2} Hz^{-1} $steradian^{-1}$. The spectral radiance can be integrated over all frequencies to yield a *total radiance* in units of erg $sec^{-1}cm^{-2}$. In SI units we have joule sec^{-1} m^{-2} Hz^{-1} $steradian^{-1}$ and watt m^{-2}. More often we use the term *spectral intensity*, measured in watt/nm for wavelength-dispersing instruments and watt/cm^{-1} for wavenumber-dispersing instruments, such as the prism and the FTS. In the case of the FTS with photometric (photon) detectors, we often use photons/cm^{-1} as the appropriate measure of radiation.

The difference between W/nm and photons/cm^{-1} is significant, because there are two types of detector. One type is a photon detector, which puts out a signal proportional to the photon density in the radiation striking it. The photomultiplier is an example. The other puts out a signal proportional to the energy density of the incident radiation. A bolometer is a detector of this type. For a given energy density, the photon density is much larger in the infrared than in the ultraviolet; and conversely, for a given photon density, the energy density is much smaller in the infrared than in the ultraviolet. In practical terms, we can see that for a given energy per unit interval (wavenumber or wavelength), a signal may saturate a photon detector if the radiation is in the infrared, while it will hardly register in the ultraviolet because there are so few photons.

To illustrate the differences between the three representations W/cm^{-1}, W/nm, and photons/cm^{-1}, we can examine the emission spectrum of a blackbody source at the temperature of the photosphere of the Sun, shown in Fig. 1.8.

The thermal radiation profile (Planck curve, given in W/nm) is illustrated in Fig. 1.8. It is familiar from radiative transfer theory as the Planck curve, and describes the energy flux as a function of wavelength:

$$B(\lambda, T) = \frac{2hc^2}{\lambda^5} \frac{1}{e^{hc/\lambda kT} - 1} \quad (\mathrm{W\ m^{-2}\ sr^{-1}\ nm^{-1}}). \tag{1.5}$$

The radiometric flux as a function of frequency is

$$B(\sigma, T) = 2hc\sigma^3 \frac{1}{e^{hc\sigma/kT} - 1} \quad (\mathrm{W\ m^{-2}\ sr^{-1}\ [cm^{-1}]^{-1}}). \tag{1.6}$$

The photometric profile or Planck curve in photons/ cm^{-1} is even less familiar

$$B(\sigma, T) = 2\pi c\sigma^2 \frac{1}{e^{hc\sigma/kT} - 1} \quad (\mathrm{photons\ m^{-2}\ sr^{-1}\ [cm^{-1}]^{-1}}). \tag{1.7}$$

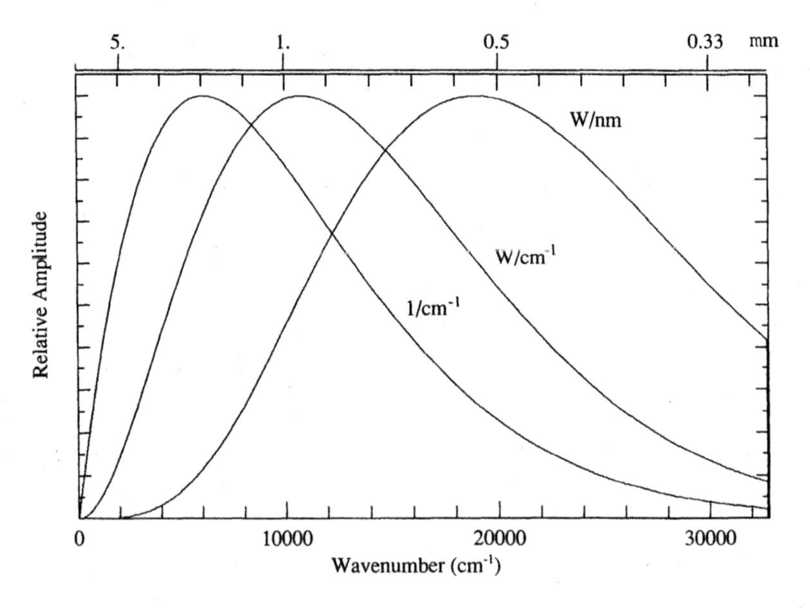

Fig. 1.8 (a) Blackbody source at 5500 K, maximum of 0.41 watts/nm at 527 nm, or 18 985 cm^{-1}. (b) Maximum of 0.30 watts/cm^{-1} at 927 nm, or 10 790 cm^{-1}. (c) Maximum of 1.8×10^{18} photons/cm^{-1}; maximum at 1642 nm, or 6089 cm^{-1}.

In these equations, h and k are Planck's and Boltzmann's constants, respectively, and T is the absolute temperature. The marked differences in shapes are due to the differences in frequency dependence: The first equation goes as σ^5, the second as σ^3, and the third as σ^2.

Standard pressure and temperature (STP) provide a reference point for all measurements of gases. They define a standard density of 1 amagat. Standard temperature is 273.15 K, and standard pressure is 1013.25 mbar, or 1.01325×10^5 N/m^{-2}, or 760 mmHg. Unfortunately, the relevant quantity in much of spectrometry is *how much* of a given species is present in a sample, which is measured in terms of the total number of particles in the line of sight, called the *column density*. The units are particles/cm^2, which can be translated into centimeter amagats, a more convenient unit for measuring column densities in a cell of known length and pressure. The conversion factor is 1 cm amagat = 2.68675×10^{19} particles/cm^2.

For absorption spectra, the spectral intensity and the column density are linked by Beers's law, which relates the absorption to the column density. For emission spectra, the amount of radiation for thin sources is linearly dependent on the number of radiating particles. Column densities can be translated into fractional abundances in parts per million by mass or volume, abbreviated ppmM or ppmV, or more typically into a concentration times a length such as parts per million meter or ppm m, although the conversion is dependent on the local atmospheric density. At sea level, 1 ppm m $\sim 2.37 \times 10^{15}$ particles cm^{-2}).

1.5 A Glance Ahead and to the Side

We hope this text serves two purposes and two communities of readers: students in chemistry and physics who are preparing for research using spectrometry, and practitioners in the field who are interested in the best methods used in the field. We first derive some basic equations that describe an idealized interferometer and then extend the description to include some unavoidable physical limitations, such as maximum path difference and finite input aperture. The next step takes us into the world of digital signal processing: sampling theory, discrete Fourier transforms, etc. Then we move into the non ideal world to look at noise and ghosts, into the practical problems involved in operating an FTS. And finally we present some applications.

We refer to our experience with the development and use of the McMath–Pierce FTS in the broader context of how to get the most from your instrument and how to understand the inevitable modifications of spectra by the instrument. The Fourier transform spectrometry community is quite large and is growing rapidly, although the high-resolution portion is small. There are many facts or rules of the game that are known to only a few individuals in the field but unknown to most of the community and consequently go unheeded. The result is that many FTS users throw away much of the information inherent in their data. Each rule is based on a mathematical expression that defines a limitation of the method, which has practical consequences in the laboratory in terms of how the data are acquired and interpreted. Look for the following rules and examples as you peruse this book. They delineate the essential steps required to set up an experiment, in the order that one ought to address them.

1. Using resolution beyond that required to determine the expected line shape adds high-frequency noise to the interferogram.
2. The size of the entrance pupil should be matched to the resolution required.
3. Three to five samples per FWHM are necessary to keep the instrumental distortion (ringing) below 0.1% of the central intensity, depending on the shape of the line (three for gaussian lines, five for lorentzian).
4. Violating the sampling theorem at any stage in the data processing introduces nonphysical features into the data. This admonition may sound obvious, but fitting minimally sampled data often violates the sampling theorem, because the algorithms compute *derivatives* of the spectrum, which require twice as many sampling points.
5. The use of apodizing functions with discontinuous derivatives creates manifold problems, so don't use them.
6. Excessive apodization may make Fourier transform spectra look like grating spectra, smooth with no ringing, but it only wastes information when used in data-processing algorithms, particularly with fitting routines.

This book has its origin in Brault's seminal paper (1985), which captured many of the lessons learned in the first decade of work on the 1-meter FTS at the McMath–Pierce Solar telescope at Kitt Peak. It is a collection of experiences and lessons learned, motivated by the desire to obtain the best possible measurements of spectral distributions of electromagnetic radiation.

2

WHY CHOOSE A FOURIER TRANSFORM SPECTROMETER?

What follows is a discussion of the merits of the Fourier transform spectrometer (FTS) and why we, the authors, each separately made it our instrument of choice.

To put our work into perspective, together we have measured or supervised measurements of a few thousand spectrum lines produced by prisms, a few more thousands from Fabry–Perot interferometers, and several million produced by diffraction gratings, and have ourselves measured tens of millions produced by Fourier transform spectrometers.

An evaluation of the usefulness of any tool must begin with an understanding of the task it is expected to perform. Our area of interest is passive spectrometry — we expect to set up a source of light and analyze its output without disturbing the source. We are practitioners of spectrometry in the region between 500 and 50 000 cm^{-1} (200 and 20 000 nm), with an emphasis on obtaining high-resolution, broadband, and low-noise spectra.

Every spectrometer has an entrance aperture, focusing optics, a dispersing element, and one or more detectors. Their comparative usefulness is characterized by the throughput (how much light passes through), chromatic resolving power (how close in energy two spectral features can be before they are indistinguishable), and free spectral range (how wide a spectral range can be viewed before two features of different wavelengths overlap in the spectral display). A block diagram might look like Fig. 2.1.

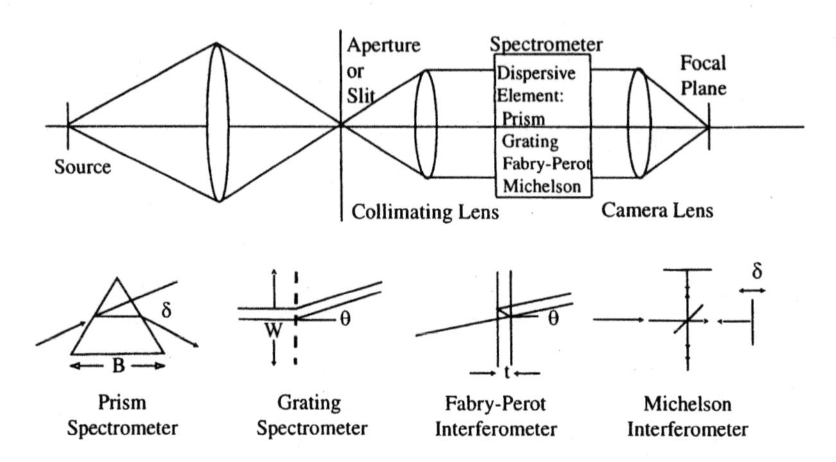

Fig. 2.1. Block diagram of spectrometer.

In this block diagram we have shown only lenses as the focusing elements, although in practice mirrors are used for almost all grating spectrometers and Michelson interferometers, and the optical path is folded back almost on itself. The FTS uses spherical mirrors at f-numbers typically between f/16 and f/50. The optical principles and practices are the same for both lenses and mirrors. In simplified terms, slice a simple positive lens in half and put a reflecting coating on the plane surface, and you have the equivalent of a concave (positive) mirror.

The job of the passive spectrometer is to gather spectral information from a source as rapidly and accurately as possible. We will consider in turn three aspects of information flow: the quantity of information per unit time, the quality of that information, and some vague sense of the cost of the information.

2.1 Quantity

The magnitude of information flow through a spectrometer may be thought of as the product of two quantities, one determined by the spectrometer optics and the other by the detector:

$$\text{information flow} = (\text{optical throughput}) \times (\text{detector acceptance}).$$

The optical throughput may be defined as the product of the area A of the entrance aperture and the solid angle Ω subtended there by the collimator, further multiplied by the optical efficiency η_0 of the system:

$$\text{optical throughput} = A\Omega\eta_0 \tag{2.1}$$

Because of its axis of symmetry, the FTS interferometer has a large entrance aperture and, consequently, a large $A\Omega$ product. A typical interferometer might have a 5-mm-diameter circular aperture.

Another aspect of the quantity of data obtained is the fact that the FTS records data at all frequencies simultaneously, a process called *multiplexing*. There is a great saving in observation time when we wish to look at many frequencies, as compared with scanning each frequency separately with a dispersive instrument such as a diffraction grating.

To determine the role of the detector on the throughput, we need to consider the mode of detection as well as the intrinsic properties of the detector. Let us combine the effects of detector quantum efficiency and the noise into a useful hybrid, the effective quantum sensitivity q, defined by:

$$q = \left[(S/N)_{\text{observed}}/(S/N)_{\text{ideal}}\right]^2 \tag{2.2}$$

where $(S/N)_{\text{ideal}}$ is the signal-to-noise ratio that would result from a perfect detector, one with unit quantum efficiency and no noise. With this concept, we define the detector acceptance as

detector acceptance = (quantum sensitivity) × (number of detectors) = qn.

The quantum sensitivity can be more usefully written as

$$q = \frac{NQ^2}{NQ + N_d}, \tag{2.3}$$

where Q is the actual quantum efficiency of the detector, N is the number of photons per measurement interval incident on the detector, and N_d is the number of detected photons per measurement interval that it would take to produce the observed detector noise (noise doesn't always come from photons!). For large signals, $NQ \gg N_d$ and we obtain $q \approx Q$, while for small signals with $NQ \ll N_d$ we have instead $q \approx (NQ/N_d)Q$, and this effective quantum efficiency depends on all three quantities, but especially strongly on the real quantum efficiency, which is not usually specified by detector manufacturers.

Finally, there are the separate but related topics of spectral coverage and free spectral range as touched upon earlier. Some spectroscopic problems can be solved by observing only a fraction of a wavenumber, while others require broad coverage, up to tens of thousands of wavenumbers. In the latter case, the amount of spectrum that can be covered without readjusting or changing components becomes a factor in the information flow. The FTS spectral coverage is limited by the beamsplitter material, beamsplitter coatings, substrate transmission, and detector sensitivity. Wavelength ratios of 5 to 1 are achievable in a single scan, and ratios of 100 to 1 are possible by switching beamsplitters or detectors or both, although the switching may not be trivial.

2.2 Quality

2.2.1 Resolution and Line Shape

Here we are concerned with the resolution and cleanness of the apparatus function, the precision of the intensity and wavenumber scales, and any possible sources of excess noise. The instrumental resolution is determined by the maximum path difference in the interfering beams. For major research instruments, this effective maximum path difference is typically 1 to 5 m, corresponding to a resolution of 0.01 to 0.002 cm^{-1}. The absolute wavenumber accuracy of any spectrum can be made to the same degree as the precision, providing there is a single standard line with which to set the wavenumber scale. Standard lines nearly equally spaced throughout the spectral region are not required to set up an accurate scale. The subject of calibration is discussed further in Chapter 9. On the other hand, many problems do not require the full resolution of such instruments. For these problems, it is useful to have variable resolution, because excess resolution reduces the signal-to-noise ratio. The FTS is especially flexible in this regard and has no equal in the ease of setting the instrumental resolution to the required value.

The accuracy in determining intensities ideally is limited only by photon statistics, but in practice there are many systematic effects that degrade performance. Some of these are apparatus function-smearing effects, which distort the shapes of spectral lines, and nonlinearity and crosstalk in detectors, which create artifacts.

One of our main concerns is with line shapes. In the past, spectroscopy has treated its two main variables very differently, being highly quantitative on the wavenumber axis but only qualitative on the intensity axis, largely because intensity measurements were difficult and unreliable. But accurate intensity information is increasingly important in many areas: modeling stellar atmospheres, unraveling complex hyperfine structure patterns, ratioing or differencing spectra to see small differential effects in the presence of large systematic effects, understanding non-voigtian line shapes, and so forth.

In measuring intensities it is necessary to take into account the apparatus or instrument function of the spectrometer, defined as the output response to a purely monochromatic input.

A major part of the value of FTS data is that a broadband interval of the spectrum can be observed in single or multiple scans with the same instrument settings and that the dispersion and the instrument line shape function are nominally the same for every spectral line no matter where it lies in the range. The FTS has an instrument function whose frequency response is essentially flat out to the end of the interferogram, where it drops suddenly to zero. All other instruments have an instrument function that changes markedly with wavelength or wavenumber. We will discuss this function in a later chapter. In the meantime, to illustrate one problem, line shape errors quantified as the decrease in peak intensity as a result of the instrument function are plotted in Fig. 2.2 as a function of resolution for both the grating spectrometer and the FTS.

If 1% line shape distortion is necessary, then an FTS with an optimum aperture as defined in Section 5.2 will require five resolution elements across a line width. In contrast, the grating with an optimum slit width will require 30 elements across a line width. The factor of 6 in required resolving power is a large part of the practical advantage of an FTS.

Wavenumber accuracy can be a large and nettlesome subject, although in the best of all possible worlds it is limited only by photon noise. Under these conditions, the uncertainty in position of a spectral line is roughly the line width divided by the product of the signal-to-noise ratio in the line and the square root of the number of samples in the line width. For example, a spectrum of N_2O taken at NSO with the 1-m FTS shows line widths of 0.01 cm^{-1} and S/N ratios of several thousand, resulting in wavenumbers with a root mean square (r.m.s.) scatter of 2×10^{-6} cm^{-1} when compared with values calculated from fitted molecular parameters. Such precision is possible though not common in modern FTS work.

2.2.2 Fixed and Variable Quantities in Experiments

There is yet another way to assess spectrometer performance, in terms of the obtainable signal-to-noise ratio. Practically, there is a trade-off among signal-to-noise ratio, spectral and spatial resolution, and measurement time, given the best electronics, detectors, and optics available.

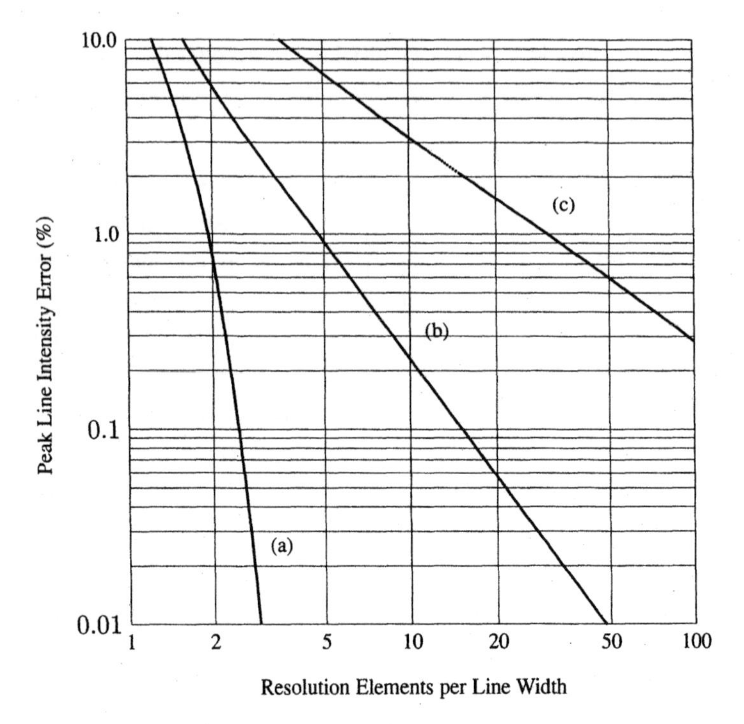

Fig. 2.2 The amplitude distortion of a gaussian line by the FTS and a grating. Curve (a) gives the limiting error for the FTS due to finite path difference alone when the aperture contribution is negligible; (b) shows the FTS error when the optimum aperture is used. Curve (c) is for a grating with an optimum slit.

All radiometric devices, including radiometers and interferometers, have common elements: an aperture of area A and solid angle Ω, and optics to channel radiation to the detector. The devices differ in their methods of spectral separation and may be compared based on the signal-to-noise ratio within a narrow spectral interval $\Delta\sigma$ that is the filter bandwidth for a radiometer or the spectral resolution width for a spectrometer.

The noise equivalent power (NEP) is the signal power for a signal-to-noise ratio of unity and is the inverse of the detectivity D

$$\mathrm{NEP}(W) = D^{-1}(W^{-1}) = \frac{\sqrt{A_D\,\Delta f}}{D*} = \frac{1}{D*}\sqrt{\frac{A_D}{T}}, \tag{2.4}$$

where A_d is the detector area, $\Delta f \sim 1/T$ is the effective bandwidth, which is determined by the dwell or integration time T at each point, and D^* is the detectivity in the detector-noise-limited regime. See Section 8.2.3 for comments on the usefulness of D^*.

The noise equivalent spectral radiance (NESR) describes the overall efficiency and throughput of the instrument:

$$\text{NESR}(\sigma) = \frac{\text{NEP}(\sigma)}{\eta_1 \eta_2(\sigma) \Delta\sigma A\Omega} = \frac{1}{\eta_1 \eta_2(\sigma) \Delta\sigma A\Omega D^*} \sqrt{\frac{A_D}{T}}, \tag{2.5}$$

where η_1 is the system efficiency and η_2 is the optical efficiency, including the transmission properties of the optical components. The spectral bandwidth is the spectral resolution $\Delta\sigma$ of the instrument, and the etendue (throughput) $A\Omega$ is the product of the collecting area and the solid angle describing the field of view.

The signal-to-noise ratio of the observation is

$$\frac{S}{N} = \frac{I(\sigma)}{\text{NESR}(\sigma)} = \frac{I(\sigma)\eta_1 \eta_2(\sigma) \Delta\sigma A\Omega}{\text{NEP}(\sigma)} \tag{2.6}$$

or, including the detector characteristics (appropriate for the infrared in the detector-noise-limited regime),

$$\frac{S}{N} = \frac{I(\sigma)\eta_1 \eta_2(\sigma) \Delta\sigma A\Omega D^* \sqrt{T}}{\sqrt{A_d}}. \tag{2.7}$$

Equation (2.7) leads to the conclusion that the best observations are obtained when the best detector is used (high D^*), the integration time is long, the condensing optics are fast (large Ω), and the bandwidth (spectral resolution) $\Delta\sigma$ is large (minimum spectral resolution).

By rearranging Eq. (2.7) we can partition the instrument performance into terms that are largely fixed and into those that are variable in the measurement design:

$$\frac{S/N}{\Delta\sigma\, \Omega \sqrt{T}} = \frac{\eta_1 \eta_2(\sigma) A I(\sigma) D^*}{\sqrt{A_d}}. \tag{2.8}$$

The right-hand side is essentially constant. System and optical efficiencies are always optimized and constrained by material properties, the aperture is as large as physically possible, the specific intensity is determined by the source and the spectral resolution required, and the detector performance is determined by its inherent properties. To gain a factor of 2 improvement requires significant investments of time and money.

On the other hand, the left-hand side is flexible in trading off one property for another. The required signal-to-noise ratio can be achieved by many different combinations of the three parameters in the denominator. These parameters are the spectral resolution, the spatial resolution or solid angle or field of view, and the observing time. We can view these parameters as axes of a three-dimensional space, as shown in Fig. 2.3.

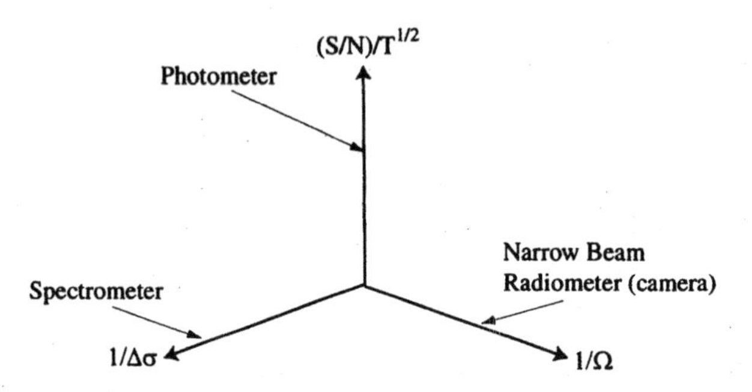

Fig. 2.3. A three-dimensional space representing the trade-off space in which instrument designs are optimized. The product of the three coordinates representing any instrument must have a fixed value determined by the right-hand side of Eq. (2.8).

Narrow-beam radiometers using camera systems emphasize angular resolution at the expense of spectral resolution and signal-to-noise ratio. In contrast, photometers trade off spectral resolution and solid angle to obtain the best possible signal-to-noise ratio in a given time interval. Finally, high spectral resolution requires compromises on the angular resolution and signal-to-noise ratio. As experimenters largely interested in high-quality spectra, we have had the luxury of practically infinite integration times and correspondingly have designed instruments with high spectral resolution and small field of view.

2.3 Cost

One concern is with the resources required to perform useful spectrometry, including not just capital outlay, but the time used in understanding and becoming familiar with the equipment, maintaining and extending it, and handling the data that justify the whole apparatus in the first place. There is a widespread feeling that grating instruments are cheap and simple and that an FTS is complex and expensive, and to some extent this is true. But the instruments being visualized when such comparisons are made are usually vastly different in power. The most challenging problems are handled not with, say, a 1-m Ebert–Fastie spectrograph with photographic recording, but with 10-m-class multiple-passed scanning gratings or an echelle crossed with a grating and having a two-dimensional spectral display, and not with a simple single Fabry–Perot etalon, but with multiple-etalon systems. However it is accomplished, high-precision spectrometry is expensive in the time of experts as well as in capital.

It is easy to ignore the cost of data reduction, but this can be a real mistake. An instrument is built once but used many times to obtain data. The natural output of an FTS after a straightforward numerical transform is a set of numbers representing the intensities on a linear scale, at a set of points equidistant in wavenumber. Computer programs exist that operate directly on such records, producing plots and lists of spectral line parameters almost automatically and making it possible to deal with spectra of quite remarkable complexity. The importance and value of such capability cannot be overemphasized. Furthermore, the required computational power, including that needed to perform the numerical transform, is readily available on personal computers.

2.4 Summary

To put these comparisons in perspective, we can take several practical cases of spectra we wish to measure and discuss which instrument we might choose.

Consider fluorescent lamps, which come with several different colors as seen by the eye — white, blue, red, etc., with no radiation outside the visible spectrum. Suppose you wanted to make a quick comparison of the color content of each. Simply look at the lamp with a hand-held prism spectroscope. To get a more precise evaluation, try a spectrometer with a 60-degree prism of base size 75 mm, a dispersion index of 50, with f/16 optics. The resolving power is 7500 (0.1 nm). It produces a single spectrum, with the visible region covering about 20 mm and no overlapping of spectral regions.

To look at the same lamps with a resolution large enough to resolve the mercury yellow lines at 577 and 579 nm, try a grating of 50-mm width used in a Littrow mounting (equal angles of incidence and diffraction) with f/5 optics – a 1/4 meter scanning monochromator, available commercially. It has a maximum theoretical resolving power of 200 000 at 500 nm. Since resolving power is most often used as the basis for comparison, remember that it is expressed as

$$R = (\text{order of interference}) \times (\text{number of grating grooves}) = mN$$

or using the grating, because $R = mN = (d \sin \theta/\lambda)(W/d) = W \sin \theta/\lambda$, the number of wavelengths that will fit into the maximum path difference between rays diffracted from opposite ends of the grating. In practice the resolving power is far less than that theoretically possible because the spectrum is observed in the first order for simplicity of data reduction, and the range of groove spacings available is limited – representative values are 300/mm, 600/mm, 1200/mm. The slit width also affects the resolving power. To put in some numbers, consider the instrument just mentioned, used in the first order with a 50-mm-wide grating having 600 grooves/mm. The theoretical resolving power is 30 000, but with a typical 5-micron slit it is reduced to 20 000. The yellow lines are easily resolved. There is no overlapping of orders because of the restricted range of visible radiation. The width of the visible spectrum is 50 mm, and a typical scan might take 3 minutes.

Now try observing the mercury green line (also present in a fluorescent lamp) with a Fabry–Perot interferometer for the purpose of examining the central line structure in detail, where a resolving power of 800 000 is needed. A Fabry–Perot interferometer with a spacing of 7 mm, a reflectance of 90%, and f/16 optics has a resolving power of 800 000 and a free spectral range of 1.5 cm^{-1}, or 0.05 nm. Here the resolving power R = (order of interference)(equivalent number of interfering beams) = $(2t/\lambda)N_R$, where N_R is the finesse, about 30 for a reflectance of 90%. In this case a narrow band filter of width 1.5 cm^{-1} is required to isolate the line from the background radiation. An auxiliary dispersing spectrometer (grating or prism) is often used for this purpose, such as the 1/4 meter monochromator described earlier.

When we wish to observe the entire lamp spectrum in great detail, including the hyperfine structure in the green line, we can use an FTS with a maximum path difference of 200 mm, which gives a resolving power of 800 000. The path difference of 200 mm is 30 times the plate separation in the Fabry–Perot interferometer, but in return there is not the same limitation on the free spectral range. The limit depends on the sampling frequency of the electronics and the speed of the moving mirror. A typical value of spectral range is 10 000 cm^{-1}, or 250 nm. A single scan might take 2 minutes. The resolution can be changed by simply changing the value of the maximum path difference. The same FTS can be changed from a low-resolution to a high-resolution spectrometer on demand, from a "quick look" instrument to observe changes in spectra with changing source conditions almost in real time to a high-resolution maximum signal-to-noise instrument. Its flexibility in this regard is unequaled.

Each of the three systems — grating, Fabry–Perot, and FTS — occupies a useful niche in the overall scheme of spectroscopy.

Broadband spectra of modest quality are most simply and cheaply obtained by the grating with photographic or CCD recording, at least in the visible and UV. This system is also the most tolerant of source intensity variation. Echelle spectrographs with array detectors bring at least an order of magnitude improvement in quantity of data gathered with a diffraction grating and in digital data processing. However, at high resolution they reproduce line shapes and positions with only modest accuracy, owing to optical aberrations and nonlinearities in dispersion. Data reduction and analysis initially require a minimum of computation to get a first look at the spectrum, but the extra computations required to convert wavelengths to wavenumbers, fit the spectral lines, and construct atlases are time consuming and full of pitfalls. A typical spectrum might consist of 20 successive echelle orders, each with a variable dispersion within an order, and a changing dispersion from order to order. The data are in wavelengths rather than wavenumbers and consequently require an extra computation to get the energies of levels. The number of samples in each spectral line must be much larger than for FTS data to get accurate fits for position, intensity, shape, width, and area, and even then the lines are always asymmetrical in shape. When constructing atlases, each order must be interpolated to the same dispersion linear in wavenumber, and then the orders must be trimmed and shifted to match each one with the preceding and succeeding ones. These computations are all doable, but not trivial.

High-resolution and compact size are the strong points of the Fabry–Perot interferometer, though it is restricted to problems that need only a small free spectral range and are tolerant of apparatus function smearing. The FTS is the system of choice in the infrared under almost any conditions (with or without a multiplex advantage) and in the visible and UV when high accuracy is required in intensity, line shape, or wavenumber.

3

THEORY OF THE IDEAL INSTRUMENT

The essential problem of spectrometry is the measurement of the intensity of light as a function of frequency or wavelength. The Fourier transform spectrometer is a multiplex instrument, meaning that the spectral information is encoded in such a manner that the intensity distribution at all frequencies is measured simultaneously by a single detector, producing an *interferogram*, as we have noticed in the previous chapters.

A simplified optical arrangement has already been shown in Fig. 1.6. A more sophisticated optical configuration is shown in Fig. 3.1. In the plane mirror configuration, one-half of the output signal is returned to the source and is lost. With retro-reflectors instead of plane mirrors, the two outputs are separable. Using the second output as well as the first doubles the system efficiency. On a more subtle level, a dual-output system provides a direct measure of the constant (DC) flux incident on the detector, which is critical for nonlinearity corrections and the removal of time-domain intensity variations. The choice of retro-reflectors stems from the fact that it is difficult to keep two plane mirrors exactly perpendicular to the optical axis as they are scanning. Optical alignment errors produce errors in the spectrum line profiles. Errors in perpendicularity need first-order corrections, while retro-reflectors require only second-order corrections.

In an interferometer, the incident light is focused onto an entrance aperture, a circular opening typically 1 to 10 mm in diameter, as shown in Fig. 1.6. Optically speaking, this aperture is the entrance pupil. The light is then collimated into plane waves, which are divided by a beamsplitter (ideally 50% transmitting and 50% reflecting) so that the two beams can travel separately through the two arms

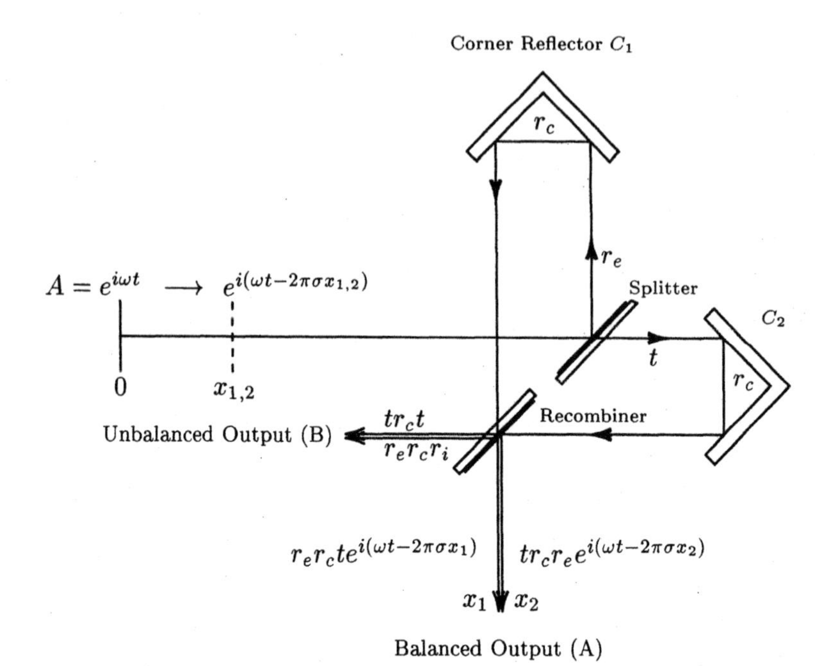

Fig. 3.1 Optical configuration for a Michelson interferometer. The symbol r_e is the external amplitude reflection coefficient at the beamsplitter and recombiner, t is the transmission, and r_c is the overall reflection coefficient of the corner reflector.

of the interferometer. The beams are reflected by mirrors and recombined by a second beamsplitter unit, the recombiner, into a single beam, which is focused onto a detector placed at the balanced output position in the figure.

In the plane of the detector the interference pattern is a set of focused circular rings called *fringes*. How many fringes are detected depends on the size of the exit pupil (the image of the entrance aperture) and on the wavenumber and difference in path length of the two interfering beams. The size of the entrance aperture is adjusted so that the central fringe at maximum path difference just fills the exit pupil where the detector is placed. The exit pupil is an image of the entrance aperture. A separate physical aperture is not needed.

With monochromatic light incident on the instrument, and when the optical pathlengths and beamsplitter phase shifts in the two arms of the interferometer are equal, then the beams interfere constructively at the detector, and the central fringe is *bright* at the balanced output. If either or both of the mirrors are moved so that the path lengths differ, then the field is bright to an extent determined by the degree of constructive or destructive interference as a result of the total optical path difference. When the mirrors are moved at constant speed, the signal at the detector alternates between light and dark in a sinusoidal fashion.

Several assumptions are implicit within the previous paragraphs that dictate the form of the following material. We have said nothing yet about how we are going to record the interferogram and take its transform. Some early instruments used bandpass filters to perform the Fourier analysis, but often the signal strength was too small to detect easily, especially in the infrared. To solve this problem the interferometer was scanned only once in a step-by-step movement of the mirrors. The interferogram was sampled at small, uniform intervals of path difference for times long enough to average out the noise. However, stepwise movement of mirrors in a Michelson interferometer is complicated to control, gives less accurate positioning, and introduces additional errors as compared to continuous scanning, as was shown by Harrison in the 1950s with his grating ruling engines. The more accurate method of data taking is to sample the interferogram while the mirrors are moving smoothly and continuously and to scan repeatedly to reduce the noise to an acceptable level, as demonstrated by the Kitt Peak instrument constructed in the 1970s.

As we shall see, the mathematical expression for the interference of the combined beams consists of two terms, a constant term and an interference term that contains all of the desired information. In some instruments the constant term is simply removed by a high-pass filter. However, there are always two recombined beams in which interference occurs as shown in Fig. 3.1, and it is possible to have two detectors and two output signals that can be combined to eliminate the constant term and double the signal amplitude, as we shall now discuss.

3.1 Equation for the Balanced Output

The *balanced* output is so called because both beams undergo one single external reflection at the beamsplitter or recombiner, and therefore produce constructive interference at the detector when the total path difference is zero. Conversely, in the *unbalanced* output, which is imaged back onto the source in a classical Michelson interferometer, one beam has a single external reflection while the other has none, and their sum has zero intensity because of a phase change of π on external reflection.

We begin by writing down the fundamental equation that describes what happens when a plane wave of monochromatic light is incident on the interferometer.

Let the incident light-wave amplitude be represented by $e^{i\omega t}$; then the amplitude of the emergent wave at the balanced output is

$$A = e^{i\omega t} r_e r_c t \left(e^{-i2\pi\sigma x_1} + e^{-i2\pi\sigma x_2}\right). \tag{3.1}$$

The emergent time-averaged intensity is the square of the amplitude

$$I = |A|^2 = 2R_e R_c T\{1 + \cos[2\pi\sigma(x_1 - x_2)]\}, \tag{3.2}$$

where $R_e = r_e^2$ and $R_c = r_c^2$ are ordinary intensity reflection coefficients and $T = t^2$ is the intensity transmission coefficient. The emergent intensity is modified by three different aspects of the interferometer. We define

$$\eta_o = \text{optical efficiency} = R_c$$

$$\eta_b = \text{beamsplitter efficiency} = 4R_e T$$

$$x = \text{path difference} = x_2 - x_1$$

and rewrite the equation as

$$I(x) = \eta_o \eta_b \left[\frac{1 + \cos(2\pi\sigma x)}{2}\right]. \tag{3.3}$$

The first term, the optical efficiency, is a simple multiplier with a maximum value of 100% when the mirrors are perfectly reflecting. The beamsplitter efficiency has a maximum value of 100% when there is no absorption and exactly half the light is reflected and half transmitted. If we use a dielectric coating so that there are no losses to absorption or scattering, $R + T = 1$ and $\eta_b = 4R(1 - R)$. Even a coating as unbalanced as $R = 0.15$ and $T = 0.85$ has an efficiency greater than 50%, and a ratio of $0.25 : 0.75$ results in 75% efficiency.

3.2 The Unbalanced Output

What about the second output? We can simply note that if no energy is lost in the beamsplitter or recombiner, the outputs are complementary, and energy not appearing at the balanced output must be at the unbalanced output. For unit input,

$$I_A + I_B = \text{constant} = \eta_o$$

$$I_A = \eta_o \eta_b \left[\frac{1 + \cos(2\pi\sigma x)}{2}\right] \tag{3.4}$$

$$I_B = \eta_o - I_A = \eta_o \eta_b \left[\frac{1 - \cos(2\pi\sigma x)}{2} \right] + \eta_o (1 - \eta_b). \tag{3.5}$$

Since both outputs contain the desired information (half the photons go in each path), we combine them appropriately by taking their difference. This has the dual advantage of doubling the signal strength and eliminating most of the constant term that introduces additive noise (see Section 8.2). There is yet another advantage to having two outputs. Their sum is a measure of the total intensity of the source, which may vary slowly in time. Using this information, the interferogram amplitude can be partially corrected for intensity variations as the interferogram is being recorded.

It should be noted that even though the preceding appears mathematically and scientifically sensible, many instruments use only one output and remove the constant term with a high-pass filter between the detector and the preamplifier. This is equivalent to subtracting the mean signal value from the interferogram, which, while normally effective, underutilizes the interferometer to avoid the complexity of a second detector. It degrades by $\sqrt{2}$ the signal-to-noise ratio the instrument is capable of achieving.

From now on, for simplicity we shall ignore the constant term and take as our measure of the output the modulation term

$$I(x) = I_A(x) - I_B(x) \sim \eta_o \eta_b \cos(2\pi\sigma x) = \eta \cos(2\pi\sigma x), \tag{3.6}$$

where we have combined the optical and beamsplitter efficiencies into a single overall efficiency $\eta = \eta_0 \eta_b$. Since it is a simple multiplier, it will be left out of the equations altogether.

3.3 From Monochromatic Light to Broadband Light

In our progression from monochromatic to broadband light, we will use the concepts and techniques of Fourier analysis, which is the process of representing an arbitrary function by a superposition of sinusoids. We will introduce each concept as required, and summarize them all in Chapter 4 with more mathematical rigor for reference purposes.

In general a source radiates more than one frequency of light, and in such cases the detector records a superposition of cosines, each one weighted according to the intensity at its given spectral frequency. For example, a spectrum consisting of two close, narrow lines of similar intensity, as shown in Fig. 3.2, produces an interferogram that looks like the familiar beat frequency phenomenon. The carrier frequency is $(\sigma_1 + \sigma_2)/2$ and the beat frequency is $(\sigma_1 - \sigma_2)$.

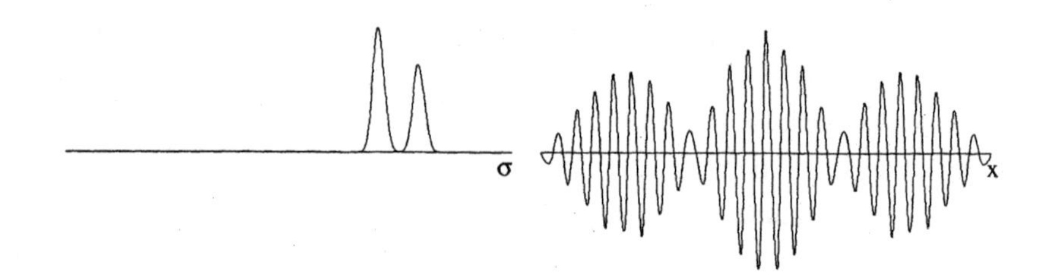

Fig. 3.2 Sodium doublet and interferogram.

Many emission sources consist of strong, apparently randomly spaced spectral lines. The interferogram of such a source displays constructive interference only near the position of zero path difference, which results in a bright central fringe, often called the white light fringe, with an amplitude proportional to N, the number of lines in the spectrum. Away from zero, the cosine wave amplitudes decrease rapidly to a value proportional to $\sqrt{N}$, as illustrated in Fig. 3.3.

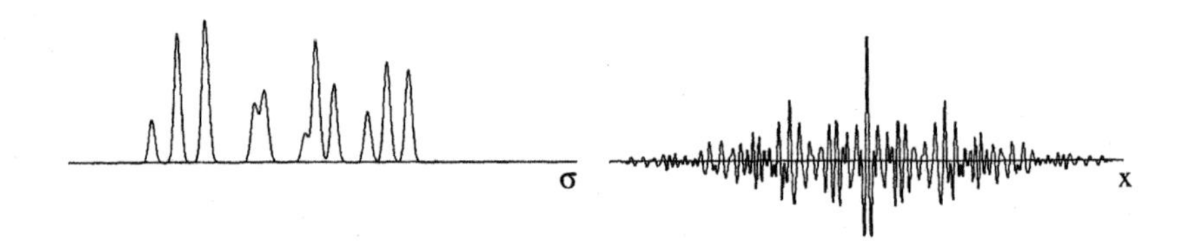

Fig. 3.3 Emission spectrum and interferogram.

Other sources emit a near-continuum of radiation with few or many absorption lines. The continuum produces only a single intense peak in the interferogram at zero path difference. Information about the absorption lines comes from the small ripples in the interferogram that extend out to large path differences. They almost look like noise, as shown in Fig. 3.4.

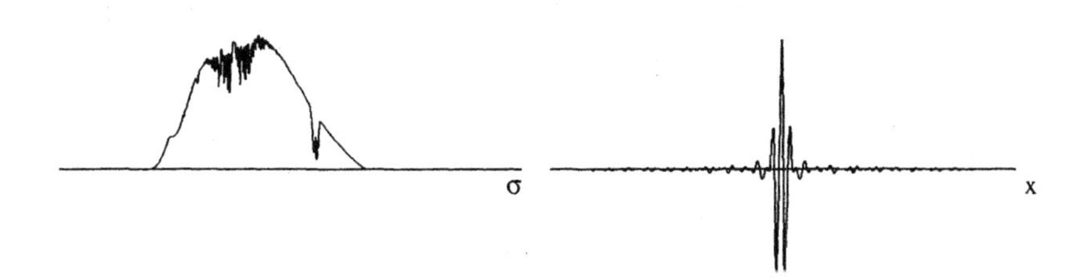

Fig. 3.1 Absorption spectrum and interferogram.

In practice, therefore, we always have a polychromatic wave rather than a monochromatic one, and we need the techniques of Fourier analysis to sort out the various frequencies. The final interferogram is a superposition of the individual interferograms for each different frequency, as we shall see.

3.3.1 Generalization to Polychromatic Light

Thus far we have assumed that the input light wave was a monochromatic wave (Eq. 3.6) of unit amplitude. We may generalize to polychromatic waves with realistic intensities by letting $B(\sigma)d\sigma$ be the energy in a spectral interval $d\sigma$ at the frequency σ and the corresponding interferogram $dI(x)$ be the energy detected at the optical path difference x:

$$dI(x) = B(\sigma)d\sigma \cos(2\pi\sigma x). \tag{3.7}$$

Integrating over the frequency variable yields the energy detected at a path difference x:

$$I(x) = \int_0^\infty B(\sigma)\cos(2\pi\sigma x)\, d\sigma. \tag{3.8}$$

The quantity we want to recover from this equation is the spectral distribution $B(\sigma)$, which we can do by taking the inverse transform of the interferogram

$$B(\sigma) = \int_0^\infty I(x)\cos(2\pi\sigma x)\, dx. \tag{3.9}$$

3.3.2 Extension to Plus and Minus Infinity

Our arguments leading up to the Fourier transform in Eq. (3.9) make sense physically, but the mathematics actually produce not only the spectrum $B(\sigma)$ but also its mirror image, $B(-\sigma)$, at negative frequencies. Remember that $\cos(2\pi\sigma x) = \cos(-2\pi\sigma x)$ and therefore $B(\sigma)$ and $B(-\sigma)$ produce identical interferograms. The negative frequencies are physically unreal. But when we consider discretely sampling the interferogram and transforming this representation of the true interferogram, the mirror spectrum at negative frequencies plays an important role. For complete symmetry in transforming back and forth from the interferogram domain to the spectral domain, we would like the integral to extend over all frequencies from minus to plus infinity. We want the interferogram and the spectrum to be symmetrical (even functions), so we need to have an expression for B that is also symmetrical. We can construct such a spectral function B_e from B, as shown in Fig. 3.5 and Eqs. (3.11) to (3.13), and change our definitions to include all frequencies:

$$B_e(\sigma) = \frac{1}{2}[B(\sigma) + B(-\sigma)] \tag{3.10}$$

$$I(x) = \int_{-\infty}^{+\infty} B_e(\sigma) \cos(2\pi\sigma x)\, d\sigma \tag{3.11}$$

$$B_e(\sigma) = \int_{-\infty}^{+\infty} I(x) \cos(2\pi\sigma x)\, dx. \tag{3.12}$$

We now have symmetric sets of functions to work with mathematically, and that will reproduce the true spectrum properly.

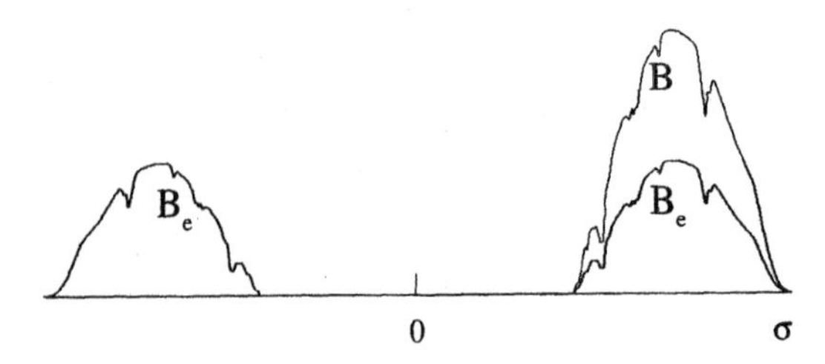

Fig. 3.5 Construction of a symmetric function B_e from an asymmetric function B.

3.4 The Fourier Transform Spectrometer as a Modulator

Now imagine that the reflectors are moved in such a way that the path difference varies linearly in time so that $x = vt$. Then the dependence of intensity on time is

$$I(t) \propto \cos(2\pi\sigma vt). \tag{3.13}$$

Since we assume the input beam is stationary (steady in spectral content and average amplitude), we can think of the FTS as a *modulator* that produces a frequency (typically audio) $f = \sigma v$ from that steady beam. Or we may think of it as a frequency multiplier that maps the light frequency σc to an audio frequency σv. The envelope of the interferogram has exactly the same shape as that of the original wave, but the "carrier" frequencies are reduced by the factor v/c and hence are easily detectable and the correct frequency *distribution* is still recoverable.

3.5 Summary

The heart of Fourier transform spectroscopy is the recognition that polychromatic spectral distributions can be determined by measuring the interferogram produced in an amplitude-division (Michelson) interferometer and then calculating the Fourier transform of the interferogram. We measure $I(x)$ and then perform mathematical operations to obtain $B_e(\sigma)$ and construct $B(\sigma)$, the desired spectrum. Figure 3.6 illustrates the process with a spectrum consisting of three lines with different widths and profiles.

While the spectroscopist's experience is largely in the spectral domain, with some practice a considerable amount of information can be inferred from the interferograms themselves.

Figure 3.6 presents three individual interferograms, their sum, and the resulting spectrum. The quasi-monochromatic frequency of each individual interferogram is indicated by the oscillation frequency in the wave train. The interferogram in (a) has the smallest frequency of the three, with a corresponding spectral line of the lowest frequency, as shown in (c). A larger number of oscillations in a unit interval of the interferogram corresponds to a higher frequency in the spectrum, as shown in the interferograms (b) and (c) and in (e).

The spectral line width determines the length of the interferogram. The smaller the line width, the longer the interferogram. The central spectral line in (e) has the longest interferogram (b).

The line shape determines the shape of the interferogram envelope. The interferogram in (a) is produced by a spectrum line with a voigtian profile having equal gaussian and lorentzian components. The interferogram in (b) is produced by a gaussian profile, and that in (c) by a lorentzian profile. Notice that the lorentzian profile interferogram has a cusp at the central maximum; in contrast, the gaussian is smooth across the central maximum, while the voigtian (a convolution of gaussian

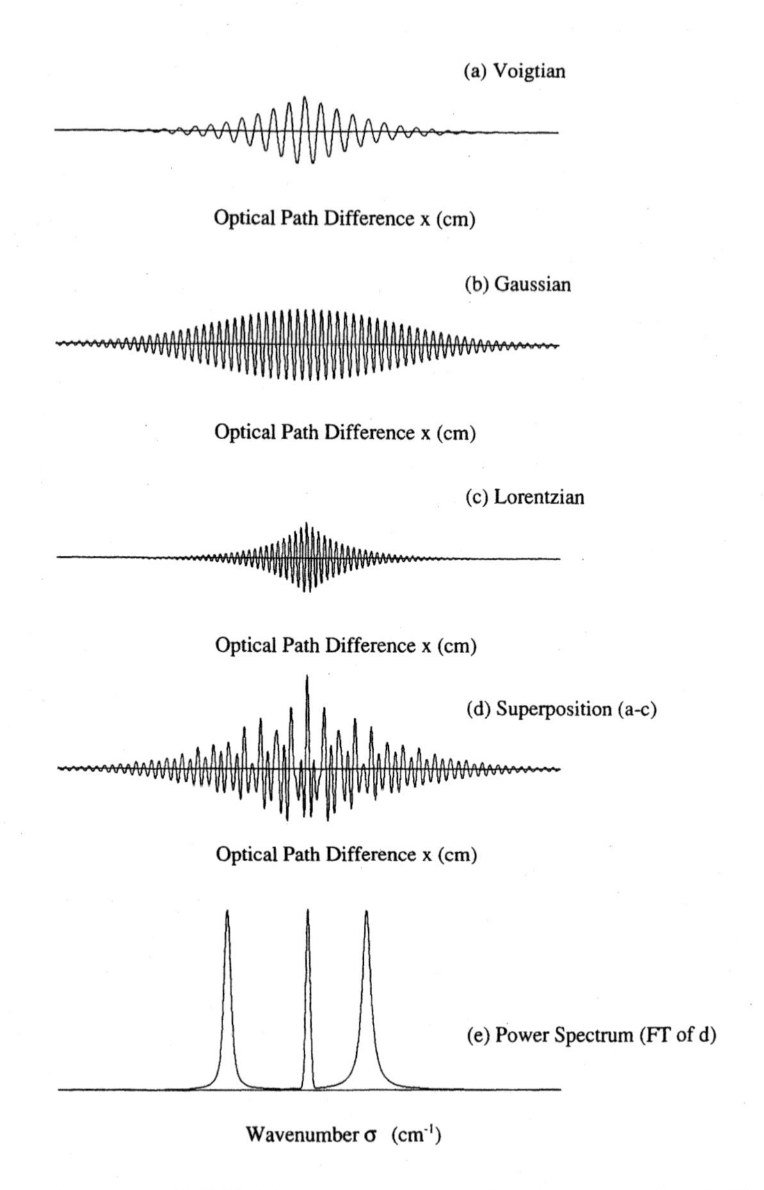

Fig. 3.6 From signal to spectrum. Individual frequency components radiated by the source (a – c). (d) Sum of the wave trains. The Fourier transform of (d) into (e) completes the cycle from wave trains to interferogram to spectrum lines.

and lorentzian profiles with the same full widths at half maximum) displays some of both constituent shapes. Also note that the lorentzian line profile (c) displays characteristic exponential wings that decay very slowly, whereas the voigtian profile decays more rapidly due to the contribution of the gaussian profile.

Of course the combined interferogram shown in (d) does not display all the foregoing features quite so distinctly, but the principles are clear and the transform of the combined individual interferograms reproduces the spectrum as exactly as if we had separate interferograms, by the principle of superposition. In many spectra all lines have roughly the same shapes and often the same widths. Then the envelope of the combined interferogram does show clearly the characteristic shape and width of the lines.

4

FOURIER ANALYSIS

At the outset of our discussion, we want to say that our points of view as physicists are not identical with those of mathematicians, especially with regard to functions that can or cannot be sampled, infinity, zeroes, and ill-posed problems. All of our computations are subject to the limitations of digital techniques and computers, where every function is digitally represented by a set of sampling points. Truly aperiodic functions do not exist. Infinite limits are too far away to be reached in practice, and negative frequencies do not exist. Zero means negligible, or impossible to pick out of the ubiquitous noise, simply not important enough to be considered. We do our best, within practical limits, to get the most accurate possible representation of the spectrum.

Sampling allows us to treat Fourier analysis computationally but at the same time introduces other complications and artifacts that place restrictions on interpretation of the transforms. These restrictions will be discussed at the appropriate places in the text.

In practice, Fourier analysis has two facets, the *construction* of a function from sinusoids (Fourier synthesis), and the *decomposition* of a function into its constituent sinusoids (Fourier decomposition). The two facets are reflected in the mathematical form of the Fourier integral. A function $f(x)$ satisfying certain mathematical conditions of continuity can be represented as a superposition of sine and cosine functions

$$f(x) = \int_{-\infty}^{\infty} F(\sigma)e^{+i2\pi\sigma x}\, d\sigma \equiv \widetilde{F}(\sigma), \tag{4.1}$$

where the function $F(\sigma)$ is termed the *Fourier transform* of $f(x)$, and can itself be expressed in a similar fashion as an integral superposition of sines and cosines:

$$F(\sigma) = \int_{-\infty}^{\infty} f(x)e^{-i2\pi\sigma x}\,dx \equiv \widetilde{f}(x), \tag{4.2}$$

4.1 Linear Systems and Superposition

The essential building block in signal analysis is the concept of a linear input-output system, a black box that can be studied by examining the output response for a variety of input excitations. Instruments, filters, and numerical algorithms can all be studied by input excitation and output response analysis. If we want to understand how a bell resonates, we ring it and listen to the frequency responses for various patterns of excitation. Few if any physical systems are actually linear, but linear approximations often constitute a sensible starting place for nonlinear analysis because within specified domains many systems are linear to first order.

In a linear system, two independent input excitations $f(x)$ and $g(x)$ generate independent outputs $F(x)$ and $G(x)$, so in the case of simultaneous excitation,

$$f(x) + g(x) \qquad \Longrightarrow \qquad F(x) + G(x), \tag{4.3}$$

which is a formal statement of the *principle of superposition*, which asserts that the presence of one excitation does not affect the response of the system to any other excitation. As an extension of this property, linear systems are also *homogeneous*, in that an excitation can be scaled linearly and the response scales identically:

$$Cf(x) \qquad \Longrightarrow \qquad CF(x). \tag{4.4}$$

This result follows because multiplication is simply successive addition.

Many linear systems are also *time invariant*, which means that a system produces a given response for a given input at any time. The response is reproducible. To be explicit, if $f(x,t)$ produces $F(x,t)$ at time t,

$$f(x,t) \qquad \Longrightarrow \qquad F(x,t), \tag{4.5a}$$

then

$$f(x,t-t_0) \qquad \Longrightarrow \qquad F(x,t-t_0) \tag{4.5b}$$

for all time delays t_0. Having this property means that no new frequencies appear in the system response — only those frequencies present in the input appear in the output.

While this may sound a little mathematical and abstract, these principles are the very basis of instrument design. Reproducibility requires that a given set of input conditions produce the same output response to within the experimental error. In Chapter 3 we constructed an interferogram by superimposing cosine waves of specified spatial frequencies and amplitudes. Such a construction assumes that the response of the interferometer is additive (so that each frequency can be treated independently) and homogeneous (so that each wave can be scaled according to the amplitude at a given frequency). Time invariance is equally important. We must go to great lengths to ensure that our instrument response either is the same for all times or can be measured and calibrated so that all measurements may be placed on a relative scale. Practically, these two conditions — linearity and time invariance — dictate that the output of the system must be related to the input of the system by a *convolution*.

4.2 A Practical Approach to Fourier's Theorem

In Chapter 3 we constructed a theoretical Fourier transform spectrometer out of a Michelson interferometer, by measuring the interference pattern produced by polychromatic radiation. This is an example of a linear system that we described by the expression (Eq. 3.8)

$$I(x) = \int_0^\infty B(\sigma) \cos(2\pi\sigma x)\, d\sigma. \tag{4.6}$$

Each frequency σ has a spectral intensity $B(\sigma)$ that defines the scale factor for that frequency and produces a cosine of spatial frequency $2\pi\sigma x$. The combined interference $I(x)$ at a position x is defined by the integral over all frequencies. The additive and homogeneous properties of a linear system are necessary conditions for this relation, and, as already mentioned, the time invariance is assumed. In Chapter 3 we asserted that the spectral distribution $B(\sigma)$ was the Fourier transform of $I(x)$. We will now justify that assumption.

Any continuous function $I(x)$ can be decomposed into the sum of an even and an odd function, and of course the reverse is also true: The appropriate combination of even and odd functions can reproduce any desired function:

$$I(x) = \frac{I(x) + I(-x)}{2} + \frac{I(x) - I(-x)}{2} = e(x) + o(x), \tag{4.7}$$

where $e(x)$ is an *even* function defined by the property $I(x) = I(-x)$ and $o(x)$ is an *odd* function defined by the relation $I(x) = -I(-x)$, as illustrated in Fig. 4.1. A significant feature of the composition or decomposition processes is that $e(x)$ and $o(x)$ must be defined over the range $-\infty \to \infty$.

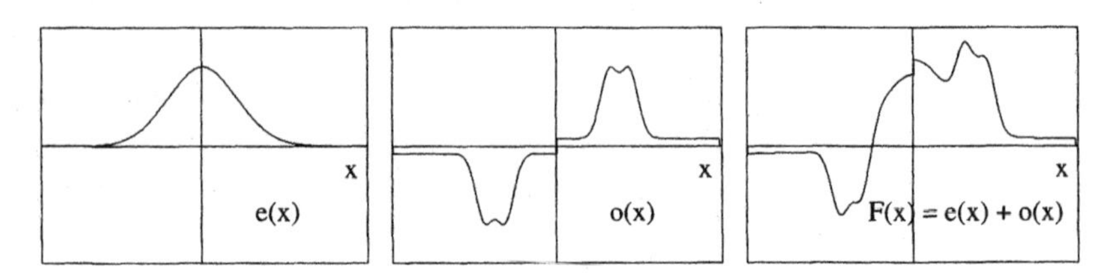

Fig. 4.1 The sum of an even (symmetric) function *e(x)* and an odd (antisymmetric) function *o(x)* is in general an asymmetric function, neither even nor odd.

An even function can be constructed from a superposition of cosine functions, an odd function can be constructed from sines, and an asymmetric function can be constructed from a combination of sines and cosines.

$$e(x) = \int_{-\infty}^{+\infty} B_e(\sigma) \cos(2\pi\sigma x)\, d\sigma \tag{4.8a}$$

$$o(x) = \int_{-\infty}^{+\infty} B_o(\sigma) \sin(2\pi\sigma x)\, d\sigma, \tag{4.8b}$$

where $B_e(\sigma)$ and $B_o(\sigma)$ specify the amplitude of each sinusoid present in the superposition. Each of them has a spatial frequency $2\pi\sigma$. An interferogram $I(x)$ is synthesized by combining sines and cosines of appropriate amplitudes:

$$I(x) = e(x)+o(x) = \int_{-\infty}^{+\infty} B_e(\sigma) \cos(2\pi\sigma x)\, d\sigma + i \int_{-\infty}^{+\infty} B_o(\sigma) \sin(2\pi\sigma x)\, d\sigma, \tag{4.9}$$

which can be rewritten as folllows by incorporating the fact that the integrals with *odd* integrands vanish (since the product of an even and odd function is always odd, and the integral of an odd function over all space vanishes):

$$\int_{-\infty}^{+\infty} B_o(\sigma) \cos(2\pi\sigma x)\, d\sigma = 0 \tag{4.10a}$$

$$\int_{-\infty}^{+\infty} B_e(\sigma)\sin(2\pi\sigma x)\,d\sigma = 0. \tag{4.10b}$$

The general decomposition of an arbitrary function into sine and cosine functions then becomes

$$I(x) = \int_{-\infty}^{+\infty} [B_e(\sigma) + B_o(\sigma)][\cos(2\pi\sigma x) + i\sin(2\pi\sigma x)]\,d\sigma, \tag{4.11}$$

or, in complex notation, with $B(\sigma) = B_e(\sigma) + iB_o(\sigma)$,

$$I(x) = \int_{-\infty}^{+\infty} B(\sigma)e^{i2\pi\sigma x}\,d\sigma. \tag{4.12}$$

This is simply a generalization of the procedure we used in Chapter 3 to superimpose cosine waves to create arbitrary polychromatic spectra.

As we have discussed, an ideal interferometer simultaneously takes incident radiation at each frequency σ and produces an interference pattern of spatial frequency $2\pi\sigma$, with an amplitude corresponding to the spectral intensity $B(\sigma)$. The interferogram is a sum of these interference patterns of different amplitudes and frequencies.

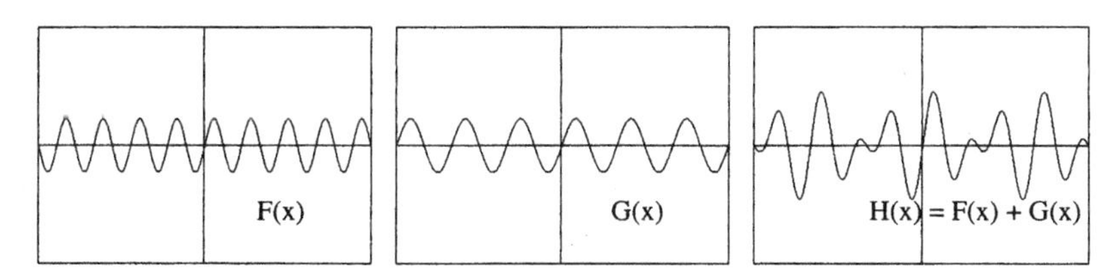

Fig. 4.2. (a) Two waves added in phase at the origin.

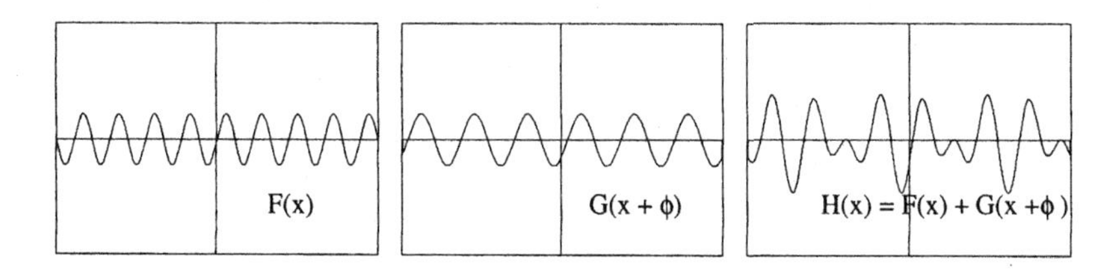

Fig. 4.2. (b) The same two waves added with a difference in phase at the origin. You may need to look carefully to see the differences, but they are there and they are significant!

In the nonideal world the preceding relationship is modified by the fact that the cosines formed for each optical frequency do not necessarily have zero phases at zero mechanical path difference but in fact have unique frequency-dependent phases. There is said to be "dispersion in the optical path." This dispersion introduces asymmetries into the interferogram, and the amplitude of radiation at each frequency cannot be recovered by a simple cosine Fourier transform. Figure 4.2 illustrates these asymmetries.

Assigning a phase (or origin) to each sinusoid introduces another degree of freedom into the spectrum—either each frequency must be assigned a unique phase so that each cosine can be shifted to the appropriate origin or we must introduce sine terms into the description of the interferogram and spectrum. Mathematically, the presence of the sine transform integral in Eqs. (4.11) and (4.12) generalizes the principle of superposition to account for any phase dispersion and permits the retrieval of spectral intensities from real-world asymmetric interferograms. Chapter 7 will provide more detailed explanations of the interplay between real and imaginary functions and phase and how we arrive at a real spectrum.

4.3 Fourier Decomposition

The inverse process is that of sorting out the frequency distribution $B(\sigma)$, given the interferogram $I(x)$ as an input. It is the process of demultiplexing the interferogram into its constituent spectral elements, or *Fourier decomposition.* To extract the function $B(\sigma)$ from Eq. (4.12), we multiply both sides by $[\cos(2\pi\sigma x) - i\sin(2\pi\sigma x)]$ and integrate over all x:

$$\int_{-\infty}^{+\infty} I(x)[\cos(2\pi\sigma x) - i\ \sin(2\pi\sigma x)]\ dx =$$

$$\int_{-\infty}^{+\infty}\int_{-\infty}^{\infty} B(\sigma)[\cos(2\pi\sigma x) + i\ \sin(2\pi\sigma x)][\cos(2\pi\sigma x) - i\ \sin(2\pi\sigma x)]\ d\sigma\ dx \tag{4.13}$$

or

$$B(\sigma) = \int_{-\infty}^{+\infty} I(x) e^{-i\ 2\pi\sigma x}\ dx. \tag{4.14}$$

But this says that the function $B(\sigma)$ is defined by a nearly identical relation to that which yielded the function $I(x)$. The similarity of the integral expressions defining $I(x)$ and $B(\sigma)$ reflects a duality in the method of describing a given phenomenon. The only difference is in the sign of the exponential. We use whichever relation is easier to visualize or use for computation.

4.4 Representation of Functions and Their Transforms

The interferogram $I(x)$ is an observed quantity and therefore by definition a real function. But because it is not perfectly symmetric, its transform $B(\sigma)$ is a complex function:

$$B(\sigma) = B_r(\sigma) + iB_i(\sigma) = |B(\sigma)|e^{[i\phi(\sigma)]}. \tag{4.15}$$

The complex function can be described as the sum of a real spectrum $B_r(\sigma)$ and an imaginary spectrum $B_i(\sigma)$, or as the product of an amplitude spectrum $|B(\sigma)|$ and a phase spectrum $\phi(\sigma)$. The descriptions are mathematically equivalent, but they draw our attention to different features of the representation. The spectrum emitted by the source is represented by a mathematically real function. However, our measurement process produces a less-than-ideal interferogram that introduces the imaginary spectrum.

The amplitude spectrum $|B(\sigma)|$ is related to the components of the complex spectrum

$$|B(\sigma)| = \sqrt{[B_r(\sigma)]^2 + [B_i(\sigma)]^2}, \tag{4.16}$$

and the phase spectrum is the angle between the imaginary and real spectra

$$\phi(\sigma) = \tan^{-1}\left[\frac{B_i(\sigma)}{B_r(\sigma)}\right]. \tag{4.17}$$

The amplitude profiles for three interferograms are shown in the left-hand column of Fig. 4.3. Our method of representing the spectrum as a complex quantity uses the Argand diagram description of complex numbers, in which a complex number is represented by its components on a unit circle. The three profiles are all real. The first is an even rectangular function, the second is an antisymmetric function, and the third is an asymmetric rectangular function. The transforms in spectral space are shown in the right-hand column; the coordinate axes are the real and imaginary amplitudes B_r and B_i. The even rectangular function transforms into a sinc ($\sin x/x$) spectrum line with only a real amplitude. The second transforms into the cosinc function lying in the imaginary plane, and the third transforms into a complex line shape function that changes phase with frequency (the rotation about the frequency axis). In addition, the asymmetric single-sided interferogram is the sum of the symmetric interferogram and the antisymmetric interferogram; correspondingly, the complex line shape function is the complex (vector) sum of the real sinc function and the imaginary cosinc function.

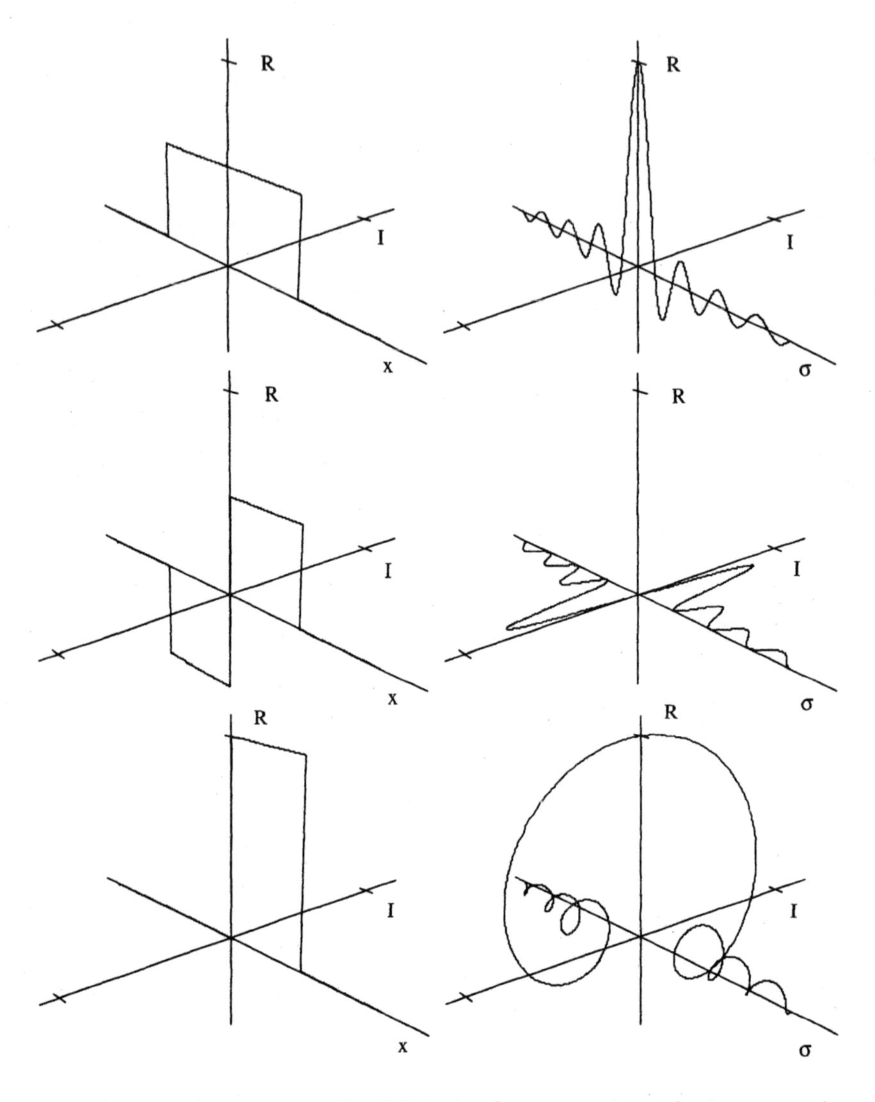

Fig. 4.3. Real and complex spectrum profiles (ILS) for interferogram envelopes of various symmetries. In the left-hand column are shown the interferogram envelopes, which are real. In the right-hand column are the corresponding transforms. Symmetric interferogram (top). The transform is real. Antisymmetric interferogram (middle). The transform is imaginary. Asymmetric single-sided interferogram (bottom). The transform is complex.

4.5. Practical Representations

A function and its transform represent two equivalent perspectives of the same information. The spectral features shown in Fig. 4.4 are fundamental examples. The finite length and varying amplitude of the interferogram determine the line width and shape. Conversely, we can go in reverse and derive the interferogram from the line shape. But this interferogram can be described as an oscillating function with an envelope. The oscillation frequency determines the central point (frequency) in transform space, while the envelope determines the shape of the spectral line. It is the envelope of the interferogram that is most important to us in determining everything but the central frequency — the intensity, shape, width, and area. Because of symmetry, we need only look at the positive x-y quadrant on a plot. We should become familiar with these representations of a spectral line, as illustrated in Fig. 4.4, because we will use each when appropriate, without further discussion. They were used earlier in Fig. 3.6.

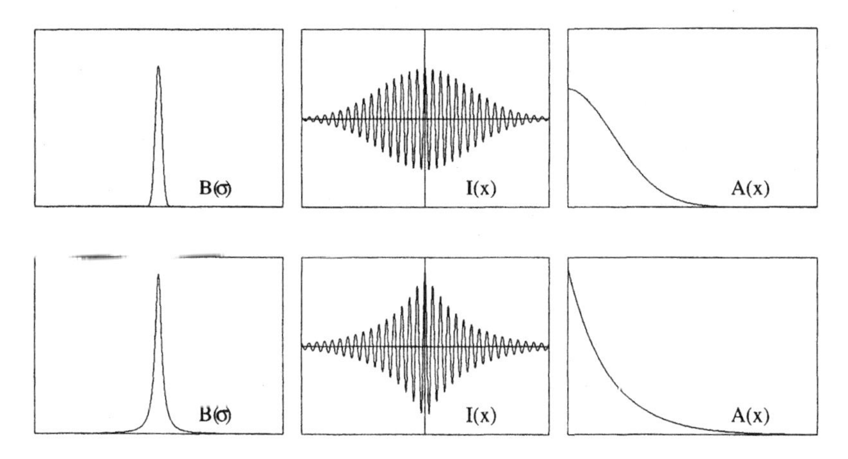

Fig. 4.4. Graphical representations of spectral lines. Gaussian profile; interferogram; amplitude of the interferogram. Lorentzian profile; interferogram; amplitude of the interferogram.

4.6 Linear Systems and Convolution

The process of observing and recording a signal does not simply duplicate the incident signal but "folds in" the properties of the instrument, even though the relationship between input and output may be a linear one. Any practical system smoothes the incident signal by taking a weighted mean over a narrow range of the independent variable. For each value of the independent variable x, the output signal $h(x)$ depends on the magnitude of the incident signal $g(x)$ at x, and its value over a range of x values. The *instrument function* (or *Green's function*) $f(x)$ describes how the input signal at x is distributed over the range of nearby x values. The observed signal at x can be described as a convolution integral

$$h(x) = \int_{-\infty}^{\infty} f(u)g(x-u)\,du \equiv f * g, \tag{4.18}$$

where we assume that the instrument function $f(u)$ does not vary with the independent variable x, so the system is both linear and *invariant* in x as well as t. The instrument function defines the resolving power of the instrument and affects all observations. The integral in Eq. (4.18) is a *convolution* integral, so called because the incident signal $g(x-u)$ is folded over with respect to the instrument function.

The geometrical optics of a scanning monochromator with wide entrance and exit slits provide a good example of the convolution operation. For a monochromatic input, the output is a convolution of two rectangular functions, in this case the slit widths. Their convolution is a triangular function, and the line shape is triangular, as shown in Fig. 4.5a, or trapezoidal when one slit is wider than the other. In this illustration we are neglecting the effects of diffraction. Think of drawing a

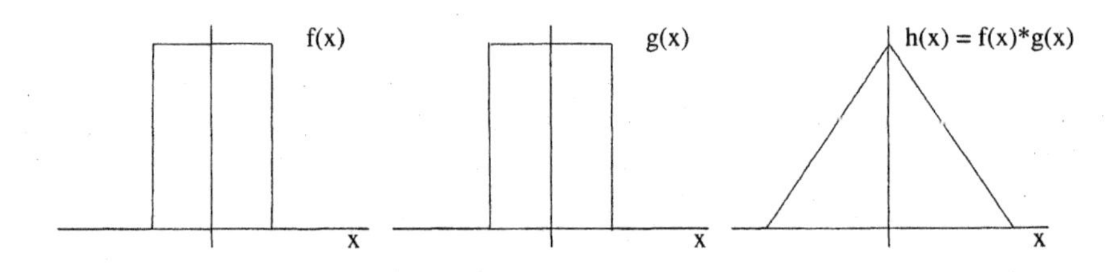

Fig. 4.5a The convolution of two rectangular functions produces a triangular function. An area normalization factor has been suppressed.

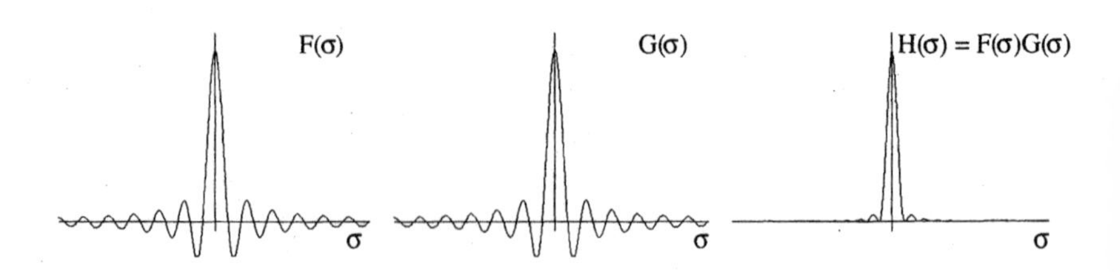

Fig. 4.5b The transform of the convolution shown in Fig. 4.5a is the product of the transforms.

rectangular aperture on each of two pieces of paper, passing one over the other, and measuring the overlapping area for each relative position.

Sometimes it is difficult to visualize the results of a convolution of two functions, as in the case where we multiply an interferogram with a rectangular or other function. We can use another relationship to help us. The Fourier transform of a convolution of two functions is the product of the transforms of the individual functions,

$$H(\sigma) = FT[h(x)] = FT[f(x) * g(x)] = F(\sigma)G(\sigma). \tag{4.19}$$

Here is an example. In Fig. 4.5a, the transform of a rect function is a sinc ($\sin x/x$) function, as illustrated in Fig. 4.5b. The convolution of two rect functions is a triangle function, whose transform is simply the product of two sinc functions, or sinc^2.

In discussing data processing and the effects of various operations like smoothing or trimming or truncating, we will work with whichever side of the equation is easier to visualize or easier to deal with mathematically. It is equally easy to visualize rect and sinc functions. But this is not so with the mathematics, where it is easy to go from a rect to a sinc but very difficult to go in the opposite direction because of the discontinuities in the rect function. When we see a sinc function we know immediately that its transform is a rect function. We do not have to crunch through the mathematics to prove it. To expand on this example, we know that an infinite wave train transforms to a delta function. A finite wavetrain is formed by multiplying the infinite wave train by a rect function. The transform of this truncated wavetrain is a convolution of a sinc function (transform of a rect function) and a delta function (transform of an infinite wavetrain). The spectrum line shape is therefore a sinc function, centered at the frequency of the wavetrain.

To implement these ideas, we note that the action of a linear invariant system (the FTS in our case) on an incident signal can be described by a *convolution of the signal with an instrument function* or by a *linear product of the signal with a transfer function.* The effect is to smooth the incident signal, which causes peaks to be spread out and reduced in amplitude. Figure 4.6 illustrates the smoothing property of a convolution. A smoothing function, in this case a gaussian filter, weights each discrete value of $g(x)$ to eliminate narrow features, as shown in Fig. 4.6. The noise is very sharp and for all practical purposes has been eliminated. The line that is half the width of the filter is markedly broadened and reduced in amplitude, while the line twice the width of the filter appears to be affected hardly at all. The original S/N ratio for both lines is 10, but it has been changed to 30 for the sharper line and to 60 for the broader line. However, the precision of position measurements has been reduced even though the S/N ratio has been increased, because the lines have been made broader.

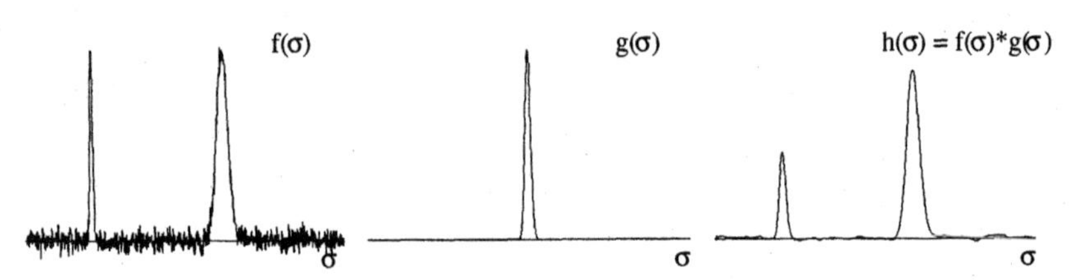

Fig. 4.6 Convolution and filtering of a noisy signal, illustrating the effects of a gaussian smoothing filter. The effect is greatest for the sharpest lines, in this case the noise.

4.6.1 Cross-Correlation

The *cross-correlation* of two complex functions $f(x)$ and $g(x)$ is

$$h(x) = f(x) \star g(x) = \int_{-\infty}^{+\infty} f^*(x) g(u+x)\, du. \tag{4.20}$$

The integral does not involve a reversal of one function (change of sign) and can be normalized to unity at the origin. Unlike the convolution integral, however, the cross-convolution integral is not commutative:

$$f(x) \star g(x) \neq g(x) \star f(x). \tag{4.21}$$

And if the functions are real, it is related to the convolution of the same functions by the expression

$$f(x) * g(x) = g(-x) \star f(x). \tag{4.22}$$

The Fourier transform of the cross-correlation of two complex functions is the cross-power spectrum

$$\Phi_{fg}(x) = FT[f(x) \star g(x)] = |F(\sigma)||G(\sigma)|, \tag{4.23}$$

which is a generalization of the power spectrum. The mean square sum of $f(x)$ and $g(x)$ integrated over all x must be a positive quantity,

$$\int_{-\infty}^{+\infty} |f(x) + g(x+u)|^2\, dx \geq 0, \tag{4.24}$$

or, after expansion of the square,

$$\int_{-\infty}^{+\infty} |f(x)|^2 dx + \int_{-\infty}^{+\infty} |g(x)|^2 dx \geq 2\left| \int_{-\infty}^{+\infty} f^*(x) g(x+u)\, dx \right|. \quad (4.25)$$

The upper limit on the magnitude of the cross-correlation of any two functions implies that cross-correlation can be used to detect features in the input signal. Equality occurs if $f(x) = g(x)$, which can happen only if the correlation function $f(x)$ is exactly the desired signal in the absence of noise.

Cross-correlation is especially useful for the comparison of data, in which the similarity or dissimilarity between a measured signal and a known signal is used to assess the presence of a desired quality in the measured signal. It is the essence of synchronous and heterodyne detection, where the measured signal is filtered or mixed with a known signal, and it is at the heart of all signal-processing algorithms. The detection process can be summarized by a single well-known but scarcely proven axiom: Optimum signal detection is achieved when the instrumental response function is chosen to achieve the maximum discrimination between the desired signal and the random noise that is ever-present in measured signals, where the maximum discrimination is defined by the maximum signal-to-noise ratio in the output signal.

Conversely, from the family of all possible instrument response functions, the optimal instrumental response function is the one that minimizes the mean square error in the process of measurement.

4.6.2 Definition of a Signal-to-Noise Ratio

The response of a system $h(x)$ to an arbitrary input $g(x)$ can be described by the convolution theorem:

$$h(x) = \int_{-\infty}^{+\infty} f^*(x) g(x-u)\, du \quad (4.26)$$

or by using Eq. (4.19) in terms of the instrument transfer function $F(\sigma)$:

$$h(x) = \int_{-\infty}^{+\infty} F^*(\sigma) G(\sigma) e^{-i2\pi\sigma x}\, d\sigma, \quad (4.27)$$

and the output power spectrum is

$$|H(\sigma)|^2 = |F(\sigma)|^2 |G(\sigma)|^2. \quad (4.28)$$

The output noise power spectrum is given by

$$|N_o(\sigma)|^2 = |F(\sigma)|^2 |N_i(\sigma)|^2, \tag{4.29}$$

where $N_i(\sigma)$ is the input noise spectrum and $N_o(\sigma)$ is the output noise spectrum, and the mean square noise is

$$< n^2(x) > = \int_{-\infty}^{+\infty} |N_o(\sigma)|^2 \, d\sigma = \int_{-\infty}^{+\infty} |F(\sigma)|^2 |N_i(\sigma)|^2 \, d\sigma, \tag{4.30}$$

where $<>$ indicates an average over all x. Consequently, the ratio of the mean square signal to the mean square noise is

$$\frac{|h(x)|^2}{< n^2(x) >} = \frac{|\int_{-\infty}^{+\infty} F^*(\sigma) G(\sigma) e^{-i2\pi\sigma x} \, d\sigma|^2}{\int_{-\infty}^{+\infty} |F(\sigma)|^2 |N_i(\sigma)|^2 \, d\sigma}. \tag{4.31}$$

Through recognizing the numerator as a cross-correlation integral we may realize that the maximum signal power is transmitted by the instrument when the instrumental response function $F(\sigma)$ is equal to the shape of the input signal $G(\sigma)$, *i.e.*, when the filter is matched to the expected signal. In all other cases, the signal-to-noise ratio will be less than the maximum possible, and hence the equality may be replaced by the inequality $\leq$ in Eq. (4.31), if a family of instrument response functions is to be considered.

4.7 Generalized Functions

4.7.1 The Dirac Delta Function

The Dirac delta-function $\delta(x)$ is a sharp spike at $x = 0$ with infinite amplitude and zero width, such that its area is normalized to 1:

$$\begin{aligned} \delta(x) &= 0 \ \ (x \neq 0) \\ \delta(x) &= 1 \ \ (x = 0) \\ \int_{-\infty}^{+\infty} \delta(x) \ \ dx &= 1 \end{aligned} \tag{4.32}$$

It has the following property when convolved with an ordinary function $f(x)$:

$$\int_{-\infty}^{+\infty} \delta(y) f(x-y) \, dy = f(x). \tag{4.33}$$

4.7.2 The Dirac Comb and Sampling of Continuous Functions

Another special function is the Dirac comb $III(x)$, which consists of an infinite series of delta functions with periodic spacing,

$$III(ax) = \frac{1}{|a|} \sum_{n=-\infty}^{+\infty} \delta(x - \frac{n}{a}). \tag{4.34}$$

The Dirac comb has two special properties, providing a bridge between continuous functions and discrete functions through its sampling property, and similarly between periodic and nonperiodic functions through its replication property. When a function $f(x)$ is *multiplied* by $III(x)$, the function is *sampled* at regular intervals, as illustrated in Fig. 4.7.

$$III(x)f(x) = \sum_{n=-\infty}^{+\infty} f(n)\delta(x - n). \tag{4.35}$$

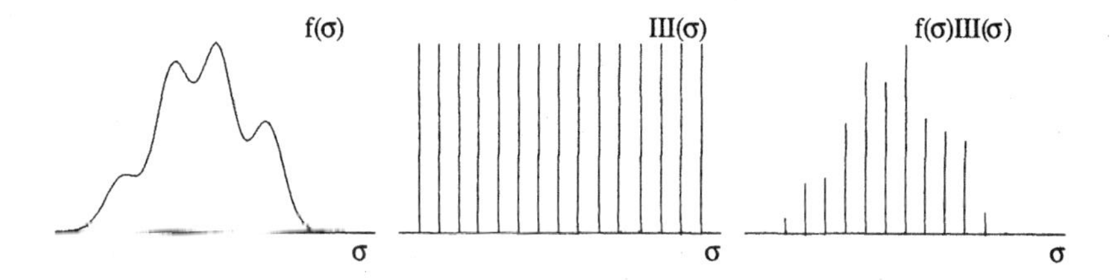

Fig. 4.7 Sampling property of the Dirac comb: spectrum (left), sampling comb (center), and sampled spectrum (right).

This property is applicable to our observed interferogram, which is a continuous function sampled at equally spaced intervals — it has been multiplied by a Dirac comb. We should note here that the transform of a comb is another comb with a different spacing, as illustrated in Section 4.9.

When a function $f(x)$ is *convoluted* with the Dirac comb, *aliasing* occurs. The function is *replicated* at unit intervals:

$$III(x) * f(x) = \sum_{n=-\infty}^{+\infty} f(x - n). \tag{4.36}$$

In common usage, the sampling theorem says that if we have a sine wave of a given frequency and wish to measure this frequency by taking the digital transform, we must sample the wave at a frequency at least twice that of the wave. By extension, if we have a periodic signal with frequency components up to some maximum value, we must sample the signal with at least twice the maximum frequency present. This minimum sampling frequency is called the *Nyquist* frequency. Another way to say it is that we need to sample a sine wave at least twice in every wavelength or period in order to measure its frequency.

Sampling and replication are intimately linked, with the Nyquist frequency setting the spacing between the spectral samples and the replication interval in the transform domain. Problems arise in signal processing when the sampling theorem is violated unintentionally. For example, in differentiating a signal, it is very common to introduce aliasing, because the differentiation process amplifies the high-frequency components of the signal. The replication process of a function at unit intervals as shown in Fig. 4.8a.

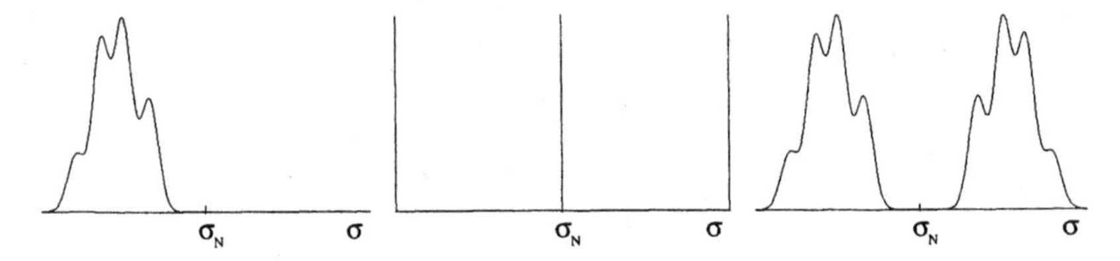

Fig. 4.8a Convolution and replication of a spectrum that is band-limited. It contains frequencies in a finite interval only. Spectrum, replication comb, and replicated spectrum. Only the first positive replica is shown. In reality the replicas extend to positive and negative infinity.

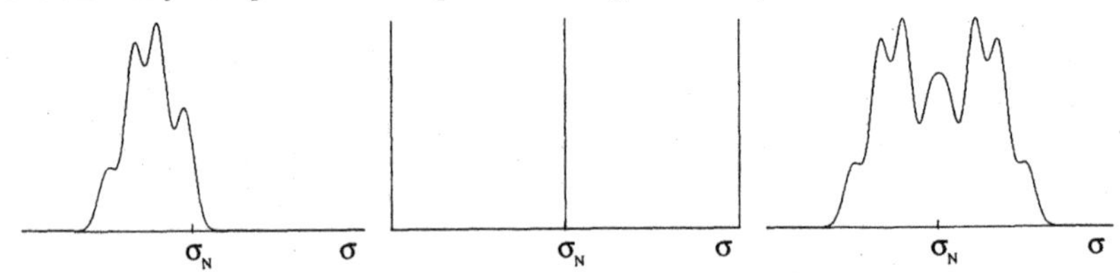

Fig. 4.8b Convolution and replication of an improperly sampled spectrum — the presence of frequencies above the Nyquist frequency causes overlap or superposition after the replication process.

Now the question arises, what if the spacing — the unit interval — of the comb in the spectral domain is smaller than the spectral profile (comb spacing too large in the interferogram domain, too small in the spectral domain)? Then the replicas are not separated as in Fig. 4.8a but overlap, as shown in Fig. 4.8b, and we say that the sampling theorem has been violated.

To summarize, all of our data are sampled rather than continuous. We can think of them as continuous functions that are multiplied by or convoluted with a Dirac comb and therefore subject to the consequences illustrated in the figures. We will say more about this at the start of Chapter 6.

4.8 The Essential Theorems

Several specific theorems play a central role in Fourier analysis, and familiarity with them provides a foundation for understanding the methods of transforming and manipulating interferograms and spectra. Each theorem will be discussed and illustrated in terms of a typical application in Fourier transform spectrometry. We will assume that there are two functions $f(x)$ and $g(x)$ in interferogram space that have transforms $F(\sigma)$ and $G(\sigma)$, respectively, in spectrum space. The Fourier transformation is indicated by the symbol $\Longleftrightarrow$.

4.8.1 Addition (Superposition)

A complex waveform is composed of the sum, or linear *superposition*, of its harmonic components in either domain:

$$f(x) + g(x) \quad \Longleftrightarrow \quad F(\sigma) + G(\sigma). \tag{4.37}$$

The superposition principle allows the creation of a complex waveform from the sum of its harmonic components. An interferogram of the sodium doublet is simply the superposition of two waves nearly identical in frequency but differing in amplitude by a factor of 2 (Fig. 4.9).

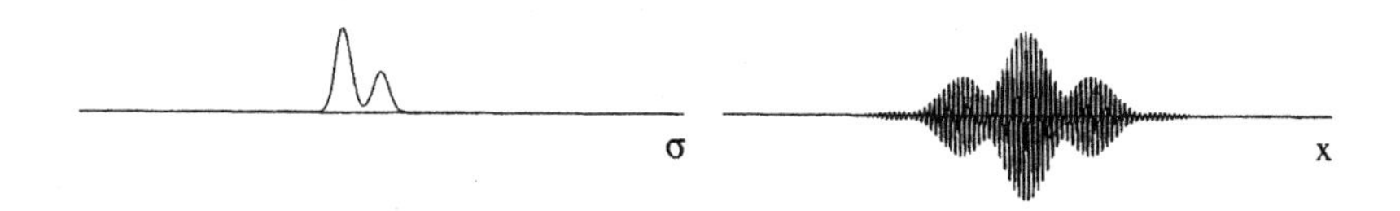

Fig. 4.9 The spectrum (left) and interferogram (right) of a spectral doublet. The beats in the interferogram amplitude have a period characteristic of the doublet splitting.

4.8.2 Similarity (Stretching)

The expansion of a wave in one domain corresponds to the compression of its transform in the other domain

$$f(ax) \quad \Longleftrightarrow \quad \frac{1}{|a|}F(\frac{\sigma}{a}). \tag{4.38}$$

The expansion or compression is scaled horizontally and vertically such that the area under the transform envelope remains constant. Compare Fig. 4.10 with Fig. 4.9. The amplitude of the doublet is doubled, but the spacing has been cut in half (a is set equal to 2). The interferogram is reduced by half in amplitude and increased by a factor of 2 in length.

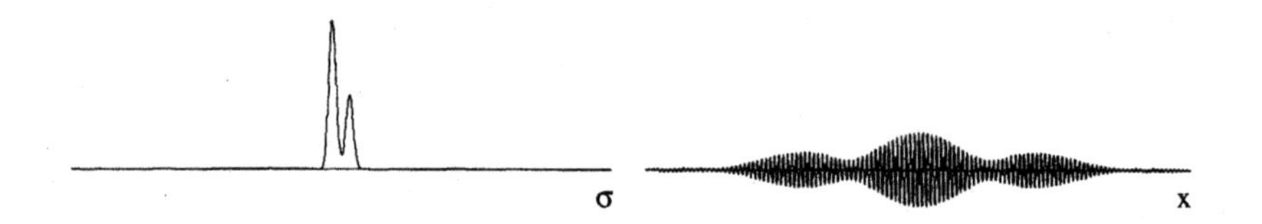

Fig. 4.10 Spectral doublet (left)compressed to half the spacing shown in Fig. 4.9, and its interferogram (right).

4.8.3 Shift

With a shift in origin, each Fourier component is delayed in phase by an amount proportional to its frequency, but the amplitude of the function, the square root of the sum of the squares of its real and imaginary parts, is not changed:

$$f(x-a) \quad \Longleftrightarrow \quad e^{-i2\pi\sigma a}F(\sigma). \tag{4.39}$$

Each unit of shift a subjects the spectrum $F(\sigma)$ to a uniform twist in phase space (Fig. 4.11). The shift theorem is the essential element of the phase correction methods that we will encounter in Chapter 7.

4.8.4 Modulation

Multiplying a function $f(x)$ by a sinusoidal wave $\cos(\omega x)$ shifts the transform $F(\sigma)$ by an amount $\pm\omega/2\pi$:

$$f(x)\cos(\omega x) \quad \Longleftrightarrow \quad \frac{1}{2}F[\sigma - \frac{\omega}{2\pi}] + \frac{1}{2}F[\sigma + \frac{\omega}{2\pi}]. \tag{4.40}$$

This is similar to the method used in radio and television to translate audio and video frequencies into radio frequencies for transmission, as shown in Fig. 4.12. We usually speak of the cosine term as the *carrier wave*, which we multiply by $f(x)$, representing the modulation signals we wish to transmit.

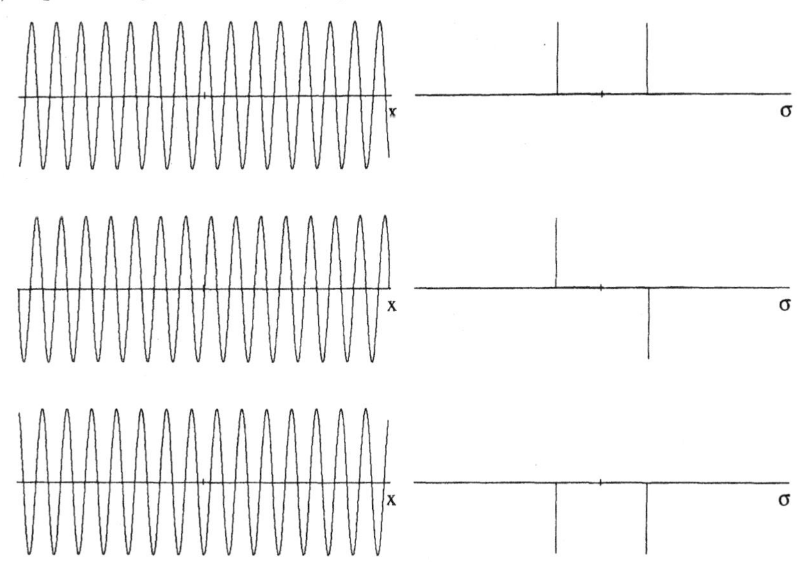

Fig. 4.11 Cosine wave and transform, sine wave and transform obtained by shifting the interferogram by 1/4 period (real to imaginary), giving 90 degrees of rotation in the complex spectrum, inverted cosine wave and transform obtained by a shifting 1/2 period, giving 180 degrees of rotation.

4.8.5 Convolution (Folding Rule)

The Fourier transform of a convolution is the product of the transforms of the individual functions:

$$\int f(z)g(x-z)\,dz \equiv f(x) * g(x) \qquad \Longleftrightarrow \qquad F(\sigma)\cdot G(\sigma) \qquad (4.41)$$

Use of the convolution theorem is extensive and implicit in many of the preceding discussions and figures, especially in the definition and modeling of instrumental functions and distortion. In most cases a complex signal chain can be decomposed into linear elements that can be modeled with a convolution. The majority of the figures in this chapter illustrate the convolution theorem: Figs. 4.5, 4.6, 4.7, 4.8, 4.12, 4.14, 4.15, and 4.17.

4.8.6 Autocorrelation

The Fourier transform of the autocorrelation function is the power spectrum, defined to be the absolute square of the spectrum $F(\sigma)$:

$$f^*(x) * f(x) \equiv \int f^*(z) f(x+z) dz \qquad \Longleftrightarrow \qquad |F(\sigma)|^2 \tag{4.42}$$

An example is shown in Fig. 4.13.

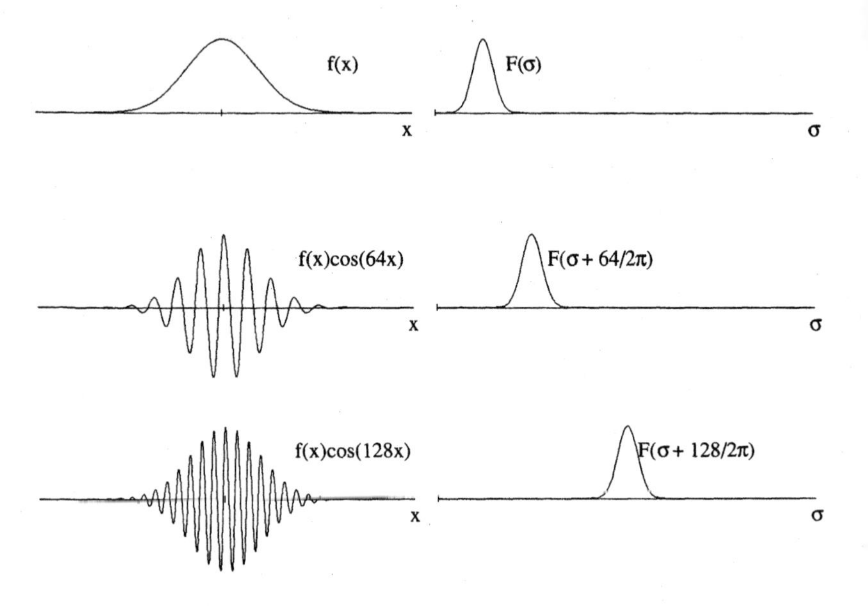

Fig. 4.12 Modulation of a function *f(x)* with cosine waves in left column, transform amplitude in right column.

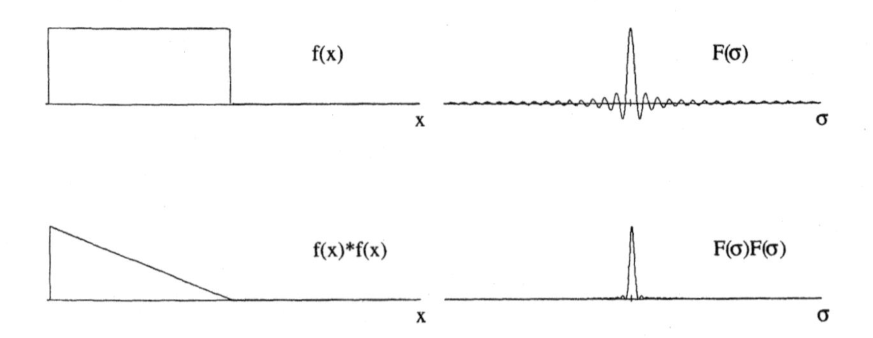

Fig. 4.13. The rect function and its autocorrelation function, with their transforms.

The transform of a rect function (boxcar) is the sinc function. The autocorrelation of a rect function is a triangular function whose transform is the sinc^2 function. To turn it around and look from the "other side of the equation," the autocorrelation of a function is the Fourier transform of the square of the transform of the function. This concept is helpful when discussing nonlinear detection of signals. In particular, higher-order correlations are often important. They appear as terms in a Taylor series expansion of a nonlinear detector system. Figure 4.14 shows a boxcar not at the origin with its autocorrelation function and a third-order (cubic) correlation function.

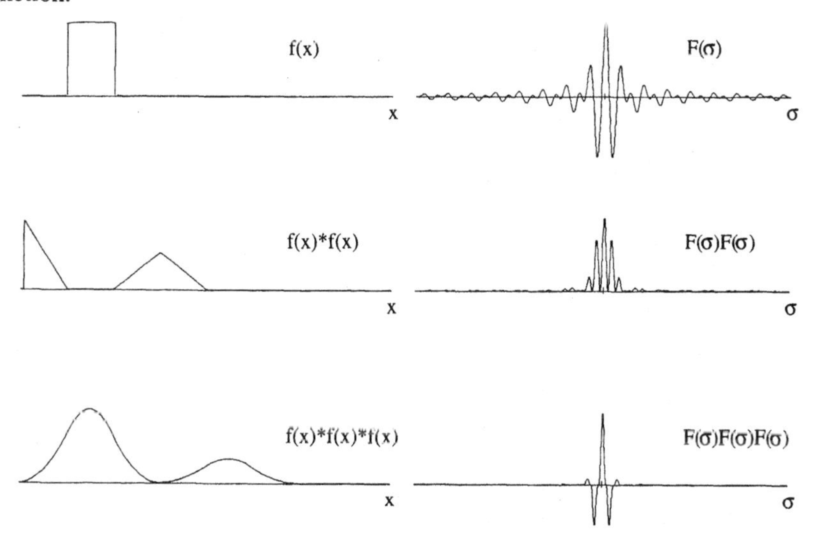

Fig. 4.14 High order correlations: A boxcar not at the origin (left) and its transform (right); autocorrelation function (left) and its transform (right), and a cubic-correlation function (left), and its transform (right).

The boxcar represents an optical filter, and the detector output might be the transform of the cubic-correlation function, owing to nonlinearities in the detector. The effective filter is not then a simple bandpass, but a more complicated one given by the cubic-correlation function, as shown in the figure.

4.8.7 Derivative

Differentiation of a function $f(x)$ is equivalent to multiplying the transform $F(\sigma)$ by $i2\pi\sigma$:

$$f'(x) \quad \Longleftrightarrow \quad i2\pi\sigma F(\sigma). \tag{4.43}$$

Consequently, differentiation of the interferogram eliminates the zero-frequency component, attenuates the low-frequency components, and enhances the high-frequency components as illustrated in Fig. 4.15. These results will have obvious consequences when a noisy signal is observed.

4.8.8 Derivatives of a Convolution Integral

Combining the convolution and derivative theorems yields

$$\frac{d}{dx}[f(x) * g(x)] = f'(x) * g(x) + f(x) * g'(x), \tag{4.44a}$$

so that

$$\frac{d}{dx}[f(x) * g(x)] \quad \Longleftrightarrow \quad i2\pi\sigma[F(\sigma)G(\sigma)], \tag{4.44b}$$

which permits the definition of useful partial differentials that are used for the measurement of signals. This result will be revisited several times in the discussions of line finding and fitting of spectra.

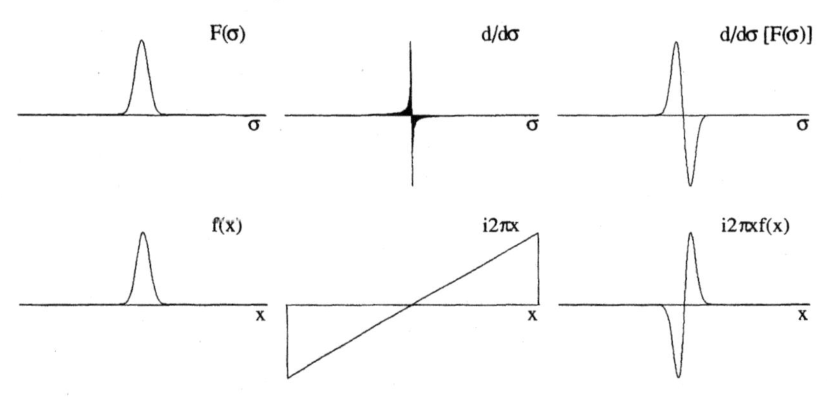

Fig. 4.15 Differentiation by convolution: (a) Direct convolution with band-limited derivative kernel is prone to approximation errors; (b) transform, product with FT*(d/dx)*, and inverse transform — while mathematically equivalent, the second approach highlights the noise-amplification property of band-limited derivative kernels.

4.8.9 Rayleigh Theorem

The integral of the absolute square of a function $f(x)$ is equal to the integral of the absolute square of its transform $F(\sigma)$:

$$\int_{-\infty}^{+\infty} |f(x)|^2 \, dx \quad \Longleftrightarrow \quad \int_{-\infty}^{+\infty} |F(\sigma)|^2 \, d\sigma, \tag{4.45}$$

which simply says that area $|f(x)|^2$ or $|F(\sigma)|^2$ is the same in both domains.

4.8.10 Power

The Rayleigh theorem may be generalized to the complex product of two functions $f(x)$ and $g(x)$:

$$\int_{-\infty}^{+\infty} f(x)g^*(x)\,dx \quad \Longleftrightarrow \quad \int_{-\infty}^{+\infty} F(\sigma)G^*(\sigma)\,d\sigma. \tag{4.46}$$

So the areas beneath the product functions fg^* and FG^* are equal.

4.8.11 Zeroth Moment

The integral of a function over all space is equal to the central ordinate of its transform

$$F(0) = \int_{-\infty}^{\infty} f(x)dx; \qquad f(0) = \int_{-\infty}^{\infty} F(\sigma)d\sigma. \tag{4.47}$$

Commonly, a characteristic *equivalent width* is defined for a function as the area of the function divided by its central ordinate, or rearranging the terms in Eq. (4.47)

$$\frac{\int_{-\infty}^{\infty} f(x)dx}{f(0)} = \frac{F(0)}{\int_{-\infty}^{\infty} F(\sigma)d\sigma} \tag{4.48}$$

which captures the essence of the duality of a function and its transform: a narrow function in one domain must include a wider spectrum than a wide function that will only require a relatively narrow spectrum.

4.9 Examples of Transform Pairs Fundamental to FTS

Six fundamental functions and their equally essential transforms are illustrated in Fig. 4.16. Where appropriate, the widths are chosen such that both the function and its transform have unit area, and consequently both are unity at the origin.

Three continuous functions and one discontinuous function reappear throughout this text: the sinc function and its transform the rect function, the gaussian function, and the lorentzian function. The convolution of the latter two functions defines the voigtian function used to model the spectrum line profiles. The convolution of the voigtian function with the transform of the rect function defines a simple model of the spectral line profile as measured by a Fourier transform spectrometer.

The Fourier transform of the sinc function sinc(σ) =sin($\pi\sigma$)/($\pi\sigma$) is the rectangular rect function $II(x)$:

$$II(x) = \int_{-\infty}^{+\infty} \frac{\sin(\pi\sigma)}{\pi\sigma} e^{+i2\pi\sigma x}\, d\sigma, \tag{4.49a}$$

where $II(x) = 1$ if $|x| < 1/2$, and $II(x) = 0$ if $|x| > 1/2$. Conversely, the transform of $II(x)$ is the sinc function

$$\text{sinc}(\sigma) = \int_{-\infty}^{+\infty} II(x) e^{-i2\pi\sigma x}\, dx. \tag{4.49b}$$

The Fourier transform of a gaussian function is itself another gaussian function:

$$e^{-\pi x^2} = \int_{-\infty}^{+\infty} e^{-\pi\sigma^2} e^{+i2\pi\sigma x} d\sigma \tag{4.50a}$$

and

$$e^{-\pi\sigma^2} = \int_{-\infty}^{+\infty} e^{-\pi x^2} e^{-i2\pi\sigma x} dx. \tag{4.50b}$$

The Fourier transform of an exponential function is the lorentzian function:

$$e^{-2|x|} = \int_{-\infty}^{+\infty} \frac{1}{1+\pi^2\sigma^2} e^{+i2\pi\sigma x} d\sigma \tag{4.51a}$$

and

$$\frac{1}{1+\pi^2\sigma^2} = \int_{-\infty}^{+\infty} e^{-2|x|} e^{-i2\pi\sigma x} dx. \tag{4.51b}$$

Notice that in contrast to the gaussian function, the lorentzian function has a smaller full width at half height when the functions are normalized to unit area, which is not the way in which the functions are typically illustrated.

Two discontinuous functions are heavily used in the discussions, the Dirac delta and the Dirac comb. The Fourier transform of $\cos(2\pi\sigma_0 x)$ is a pair of Dirac delta functions at the frequencies $\pm\sigma_0$:

$$\cos(2\pi\sigma_0 x) = \int_{-\infty}^{+\infty} \frac{1}{2}[\delta(\sigma-\sigma_0) + \delta(\sigma+\sigma_0)] e^{+i2\pi\sigma x}\, d\sigma \tag{4.52a}$$

and

$$\frac{1}{2}[\delta(\sigma-\sigma_0) + \delta(\sigma+\sigma_0)] = \int_{-\infty}^{+\infty} \cos(2\pi\sigma_0 x) e^{-i2\pi\sigma x}\, dx. \tag{4.52b}$$

The Fourier transform of a Dirac comb $III(ax)$ is itself another Dirac comb $III(\frac{\sigma}{a})$:

$$III(ax) = \frac{1}{|a|} \sum_{n=-\infty}^{+\infty} \delta(x - \frac{n}{a}) = \int_{-\infty}^{+\infty} III(\frac{\sigma}{a}) e^{+i2\pi\sigma x}\, d\sigma \tag{4.53a}$$

and

$$III(\sigma/a) = a \sum_{n=-\infty}^{+\infty} \delta(\sigma - an) = \int_{-\infty}^{+\infty} III(ax) e^{-i2\pi\sigma x}\, dx; \tag{4.53b}$$

the similarity theorem may be applied to the Dirac comb as indicated: Doubling the spectral resolution requires sampling the interferogram over twice the distance at the same interval.

Lastly, most of the considerations in this text will focus on obtaining the best possible result from a linear system given a nonideal input that contains noise. For the most part we will consider the consequences of stationary *white* noise, noise that is time invariant and possesses a uniform spectral distribution. The Fourier transform of white noise is white noise; other cases will be addressed in Chapter 8.

To repeat what we said earlier in discussing data processing and the effects of various operations, we will work with whichever side of the equation is easier to visualize or to deal with mathematically. It is easy to visualize sinc and rect functions, but it is difficult mathematically to go from sinc to rect because of the latter's discontinuities. When we see a sinc function we know immediately that its transform is a sinusoid multiplied by a rect function, and we need not crunch through the mathematics to prove it. As a more general illustration, it is easy to see the effect of multiplying two functions, but it is not so easy to visualize the convolution of two functions.

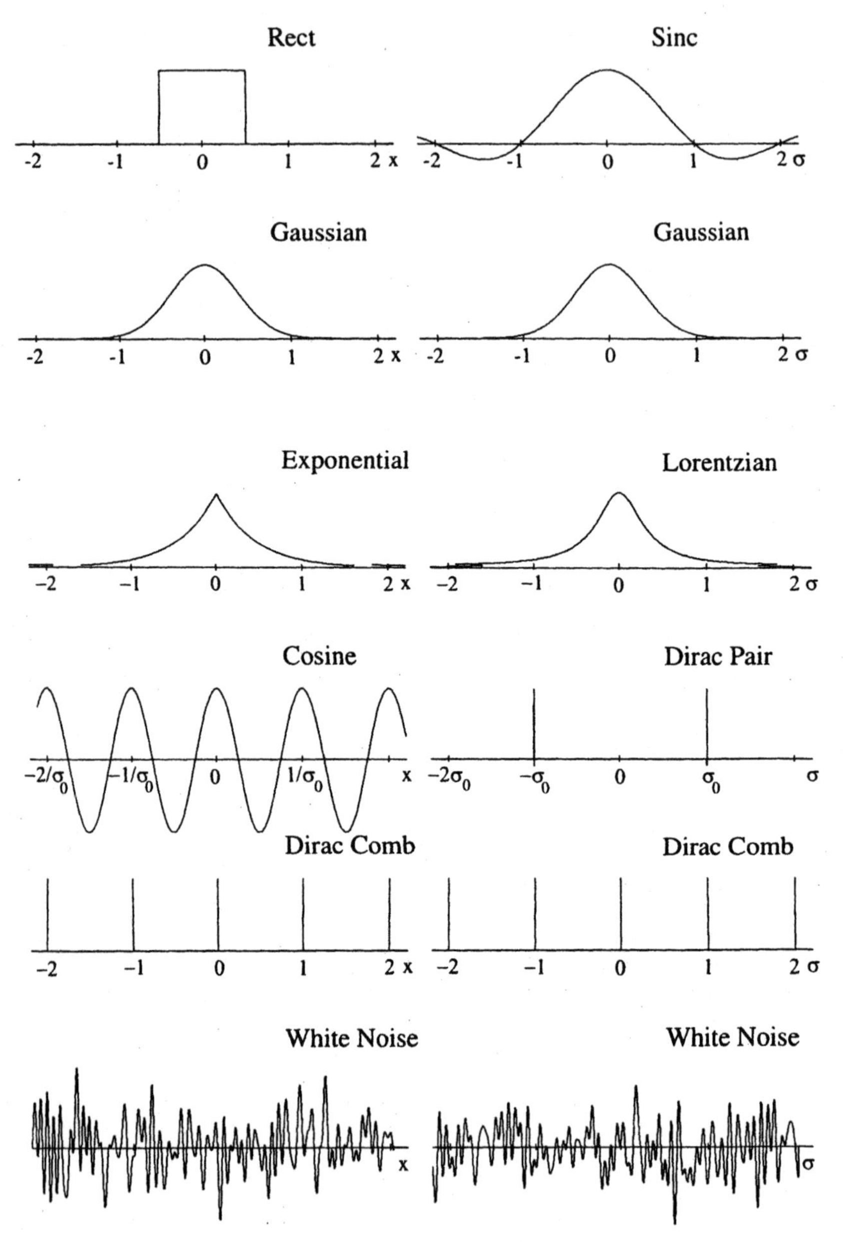

Fig. 4.16 Rectangle function and transform (sinc function); gaussian function and transform (gaussian function); exponential function and transform (lorentzian function); cosine function and transform (delta function); Dirac comb and transform (Dirac comb); white noise and transform (white noise).

5

NONIDEAL (REAL-WORLD) INTERFEROGRAMS

In our nonideal world, the spectrum is a real but not symmetric function, and the interferogram is not a real, symmetric function because of instrumental limitations and imperfections, and because the source radiation is never truly stationary (invariant in time in both spectral composition and intensity). To recover the true frequency spectrum we must find a way to compensate for the limitations of the spectrometer. In our discussion we will move back and forth between the interferogram and the spectrum in order to elucidate the points under discussion and to construct a framework for understanding how to compensate for instrumental limitations. The problems associated with one-sided interferograms will be taken up when discussing two other topics, phase correction and noise.

We will use the *instrumental function* (or instrumental line shape — ILS) as a measure of the nonideal nature of the instrument. It is defined as the shape or profile of the spectral line resulting from a truly monochromatic input. The ideal profile is a delta function. In practice it is a function much different from the ideal but nevertheless one that is mathematically representable to a high degree of accuracy.

In previous sections we considered an interferogram produced by an interferometer with an infinite path difference and an ideal optical system. A practical interferometer has a finite optical path difference, and the optics have a finite size even though they may be perfectly designed and constructed. The effect of these limitations is to produce an instrumental profile that deviates from the ideal case of a delta function. The interferogram of monochromatic radiation when transformed does not produce a spectral line of zero width, but a delta function convolved with both a sinc function and a rectangular function. The reason for this is that the finite travel of the interferometer introduces a rectangular function that multiplies the interferogram, or correspondingly a sinc function that smears (convolutes) the *spectrum*. Also, the finite size of the entrance aperture introduces a sinc function that multiplies the *interferogram*, which in turn convolutes the spectrum with a rectangular function. We now consider the details.

5.1 Finite Path Difference

To see the effects of deviations from idealized conditions, consider monochromatic light, where the interferogram is initially described by the expression

$$I(x) = \cos(2\pi\sigma_0 x). \tag{5.1}$$

If the path difference were extended to infinity, recovery of the spectrum would be perfect. The spectral distribution would be a δ-function at $\sigma = \sigma_0$:

$$B(\sigma) = \delta(\sigma - \sigma_0). \tag{5.2}$$

Since that is impractical, consider a finite maximum path difference $-L \leq x \leq +L$. This is equivalent to taking an infinitely long interferogram and then multiplying it by the rectangular function,

$$I_{\text{obs}}(x) = I(x) \cdot \Pi\left(\frac{x}{2L}\right). \tag{5.3}$$

In the spectral domain, the equivalent operation is convolution with a sinc function,

$$B_{\text{obs}}(\sigma) = B(\sigma) * 2L\,\text{sinc}(2L\sigma), \tag{5.4}$$

so the instrumental function $O(\sigma)$ is now a sinc function,

$$O(\sigma) = 2L\,\text{sinc}(2L\sigma). \tag{5.5}$$

The true line profile of infinitesimal width becomes a line of finite width, with a full width at half maximum (FWHM) of

$$\text{FWHM} = 1.207\delta\sigma = 1.207\left(\frac{1}{2L}\right) \tag{5.6}$$

We take as the width of a resolution element $\delta\sigma$, also called the *resolution width*, the distance between statistically independent samples in the spectrum $1/2L$. This resolution width is independent of σ. The maximum path difference L is sometimes referred to as the *Maximum Optical Path Difference*, or MOPD, and the resolution element is then 0.5/MOPD. The corresponding resolving power R is defined by

$$R \equiv \frac{\sigma}{\delta\sigma} = 2L\sigma, \tag{5.7}$$

and the shape is described by the instrumental function shown in Fig. 5.1.

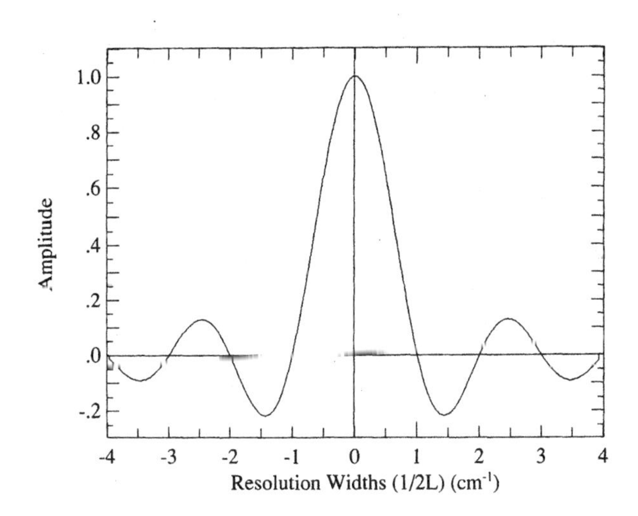

Fig. 5.1 Instrumental line shape owing to a finite length of mirror travel.

This line shape is peculiar to Fourier transform spectroscopy and appears to be nonphysical, since the negative sidelobes contradict our usual concept of light intensities. The phenomenon of sidelobes is called *ringing*. It can be suppressed by apodizing the interferogram, a process of smoothing the end discontinuities. A better way is to fit the spectrum with a model line shape incorporating the ringing in the model of the line. See Chapter 6. With broad spectral lines the interferogram is effectively self-apodizing, because its amplitude falls rapidly at large path differences, and ringing is not observed. Another way to express this is to say that ringing is apparent only when the line width is comparable to or smaller than the width of a *resolution element* of the instrument.

A quantitative discussion is given in Section 5.3, where path-length effects are combined with aperture effects in a discussion of the finite instrumental function.

5.2 Finite Entrance Aperture Size

An ideal interferogram is produced by the interference of two plane waves moving in the same direction, a condition obtainable only if the radiation source (entrance aperture) is a point located at the focus of the collimating mirror or lens. But to get a measurable amount of radiation through the spectrometer, a finite, and often, large entrance aperture is needed. Then the collimated light traverses the spectrometer at a range of angles off-axis, and the path differences extend over a range of values rather than having a single value for on-axis light. The off-axis path difference for light at an angle α is $x \cos\alpha$, and the interference fringe intensity for a particular wavenumber at a given path difference x and angle α is

$$dI = \cos(2\pi\sigma x \cos\alpha)\, d\Omega, \tag{5.8}$$

where $d\Omega$ is a small increment of solid angle at α . With the small-angle approximation $\cos\alpha \approx 1 - \alpha^2/2$, the expression becomes

$$dI = \cos(2\pi\sigma x\left[1 - \frac{\alpha^2}{2}\right])\, d\Omega. \tag{5.9}$$

We can relate the solid angle of a circular aperture at the focus of the collimating mirror to the angle α by simply noting that $\Omega = \pi\alpha^2$, since

$$\frac{\alpha^2}{2} = \frac{\Omega}{2\pi}. \tag{5.10}$$

Integrating over the aperture, which extends to Ω_m,

$$I(x) = \int_0^{\Omega_m} dI(x) = \int_0^{\Omega_m} \cos\left\{2\pi\sigma x\left[1 - \frac{\Omega}{2\pi}\right]\right\} d\Omega \tag{5.11}$$

$$I(x) = \Omega_m \operatorname{sinc} \frac{\sigma x \Omega_m}{2\pi} \cos\left\{2\pi\sigma x\left[1 - \frac{\Omega_m}{4\pi}\right]\right\}. \tag{5.12}$$

The finite aperture has produced two effects. The first is a scale change, which is easily accounted for by using $x(1 - \Omega/4\pi)$ for path difference and $\sigma(1 - \Omega/4\pi)$ for wavenumber in all calculations. Is this scale change significant? Yes: The contraction is 1 part in 10^6 for a 2-mm-diameter entrance aperture with collimating optics of 500-mm focal length. The correction is 0.010 cm^{-1} at 10 000 cm^{-1}.

The second effect is to multiply the envelope of the interferogram with a sinc function; and since σ appears in its argument, the effect varies with wavenumber. The presence of the sinc function suggests that the interferogram should be terminated at a finite path difference before the sinc function goes negative, since using the interferogram past this point actually *reduces* the signal while at the same time increases the noise, as we shall see later. The sinc function goes negative when

$$\frac{\sigma x \Omega_m}{2\pi} > 1. \tag{5.13}$$

Most commonly, the physical problem being studied requires a certain minimum resolving power R and as large an entrance aperture as possible. So let us assume some desired $x_{\max} = L$ such that $R = \sigma/\delta\sigma = 2L\sigma$ is adequate resolving power. Further, define $k = R\Omega_m/4\pi$ so that taking $k < 1$ restricts us to the initial region, where the sinc function is always positive.

To get a physical feeling for what is happening and what k represents, let us take the light emerging from the interferometer and project it onto a screen placed at the focal plane of the projecting lens, as shown in Fig. 5.2. To perform this observation we must use a sharp line source such as a slightly expanded laser striking a diffusing screen, a large entrance aperture, and a projection screen.

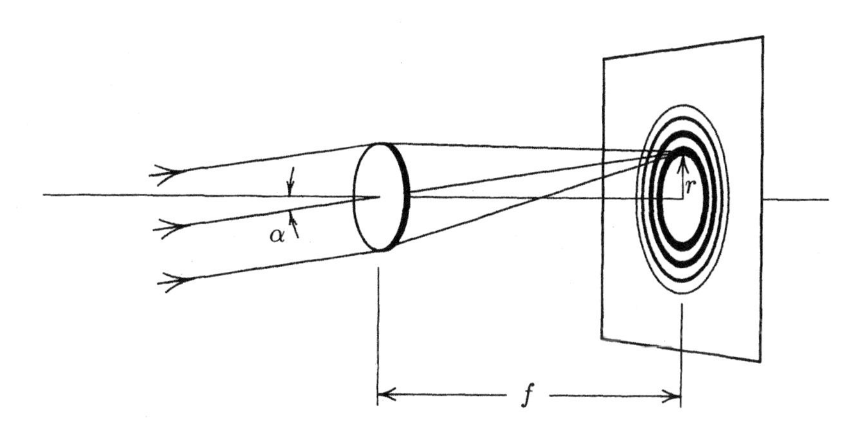

Fig. 5.2 Interferometer fringes at the focal plane.

The light coming through the interferometer at an angle α comes to a focus at a distance $r = f \sin \alpha \approx f\alpha$ from the axis. *Angles* in the interferometer are mapped into *radii* on the screen. And since we have already seen that the path difference varies as $x \cos \alpha$, we may expect to find intensity variations with r similar to those observed on the axis by varying x.

In fact,

$$x \cos \alpha \approx x\left(1 - \frac{\alpha^2}{2}\right) \approx x\left(1 - \frac{r^2}{2f^2}\right), \tag{5.14}$$

so the effective path difference for a point on the screen varies as r^2. Because of axial symmetry we therefore observe circular fringes whose spacing is described by

$$\frac{xr^2}{2f^2} = m\lambda, \qquad \text{m an integer} \tag{5.15}$$

or

$$r^2 = \left(\frac{2f^2}{x\sigma}\right)m. \tag{5.16}$$

Visually these fringes appear as thick rings widely spaced near the center, changing into thinner rings closely spaced far from the center, as shown in Fig. 5.2. To go back to the solid angle as variable, it is

$$\Omega = \pi\alpha^2 = \frac{\pi r^2}{f^2} = \frac{2\pi}{x_o\sigma_o}m = \frac{4\pi}{R}m, \tag{5.17}$$

so m is identical to the multiplier k defined earlier; *i.e.*, k tells us *how many fringes* are being accepted at the output of the interferometer. The exit aperture (exit pupil) is usually the same size as the entrance aperture, because the collimator and camera mirrors commonly have the same focal length.

To describe the effect in qualitative terms, at small path differences the exit pupil accepts a small part of a single fringe. As the path difference gets larger, the fringe width gets smaller, and the aperture accepts a larger fraction of the fringe. At a still larger path difference, an entire ring is accepted by the aperture. Past this point, more than one ring is accepted, and as a result the signal is reduced rather than increased at larger path differences, while the noise is increased.

There is no unique optimum condition on aperture size that can be applied to a spectrum containing a range of wavelengths, since both Ω_m and R depend on σ. So how shall we choose Ω_m and the related aperture size in practice? The most commonly used condition is to ask for *maximum fringe amplitude* for the *largest wavenumber* at the *longest path difference*. Going back to our original equation, the fringe amplitude is proportional to

$$\Omega_m \,\text{sinc}\, \frac{\sigma_m L\Omega_m}{2\pi} = \Omega_m \frac{\sin(\sigma_m L\Omega_m/2)}{(\sigma_m L\Omega_m/2)} = \frac{2}{\sigma_m L}\sin\frac{\sigma_m L\Omega_m}{2}. \tag{5.18}$$

The condition then reduces to maximizing the sine function by setting its argument equal to $\pi/2$:

$$\frac{\sigma_m L\Omega_m}{2} = \frac{\pi}{2}; \qquad \Omega_m = \frac{\pi}{\sigma_m L} = \frac{2\pi}{R_m} = \frac{\pi {r_m}^2}{f^2}, \tag{5.19}$$

which leads to $k = 1/2$. It may be that mechanical or optical constraints in the design of a real instrument will limit us to smaller values of k than this, especially at low resolution. And note that *largest wavenumber* really means *largest useful wavenumber* and not necessarily the highest wavenumber in the free spectral range. A plot of the fringe amplitude as a function of path difference for several values of k is given in Fig. 5.3.

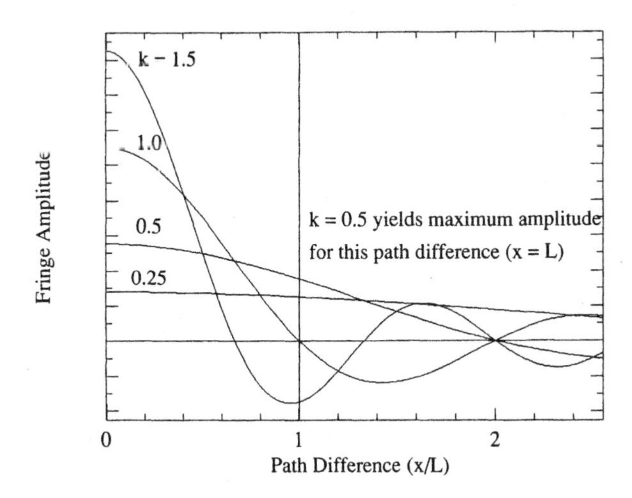

Fig. 5.3 Fringe amplitude as a function of path difference for differing entrance aperture sizes.

A very convenient expression for choosing the diameter of the optimum aperture is simply

$$d = f\sqrt{\frac{8}{R}}.$$

For example, a maximum wavenumber of 10 000 cm^{-1}, or 1 000 nm (in vacuum), a maximum path difference of 200 mm (hence a resolving limit of 0.025 cm^{-1}, or 0.0025 nm, a resolving power of 400 000), and a focal length of 500 mm, give an aperture diameter of 2.24 mm.

5.3 Finite Instrumental Function and Its Properties

The instrumental limitations already discussed can be combined into an *instrumental function*, the instrumental response to a monochromatic input signal, as described earlier in this chapter. Since most experiments require a certain minimum resolving power R and as large an entrance aperture as possible, we need to develop expressions for the output line widths and how they depend on the instrumental function.

The ideal cosine wave interferogram of a monochromatic line is multiplied by two factors, one for the finite-sized aperture and the other for the finite maximum path difference. The results, indicated by $\widetilde{O}$, may be called the instrumental function.

$$\widetilde{O} = \left[\Omega_m \operatorname{sinc} \frac{\sigma_o x \Omega_m}{2\pi}\right] \cdot \Pi\left(\frac{x}{2L}\right). \tag{5.20}$$

The instrumental line shape function (the spectral convoluting function, the Fourier transform of the foregoing function) becomes

$$O(\sigma) = [\Omega_m \frac{2\pi}{\sigma_o \Omega_m} \Pi(\frac{2\pi\sigma}{\sigma_o \Omega_m})] * [2L \operatorname{sinc}(2L\sigma)]. \tag{5.21}$$

The observed spectrum is the convolution of the ideal spectrum with this instrumental lineshape function

$$B_{\text{obs}}(\sigma) = B(\sigma) * O(\sigma). \tag{5.22}$$

The instrumental function introduces two effects into the spectrum. The spectrum is smeared with sinc $2L\sigma$, as before, and the spectrum is also smeared with a rectangular function of width $w = \sigma_o \Omega_m / 2\pi$, which equals σ_m / R for $k = 1/2$. In this case the rectangle is exactly one resolution width wide at σ_m. Refer to Fig. 5.1 and the discussion thereof. If we maintain this condition, the dominant effect is due to finite path difference when the line being observed is underresolved, but as the resolution increases the finite aperture contribution takes over. The maximum error in amplitude is always at the line center in either case, but the error due to finite path difference may cause ringing, as mentioned, while the effect of the finite aperture is simply to broaden and weaken the line.

The essential issue is that an instrument with a specified spectral resolution width $1/2L$ cannot resolve a spectrum line of a similar width without significant line shape distortion resulting from the instrumental line shape function. Part of the distortion, usually the smaller part, is a result of a finite aperture size. It is usually neglected in most discussions. The larger part is a result of the finite path length L. Let's illustrate these effects for an instrument with a spectral resolution of 20-mK (1 mK = 0.001 cm^{-1}). Figure 5.4 shows the effects of convolving a 20-mK sinc function with gaussian lines of equal peak height and widths 10, 20, and 40 mK (1/2, 1, and 2 times the instrumental resolution width $1/2L$).

When the spectral line width is smaller than the instrumental width, the ringing is pronounced. When the widths are equal, it is noticeably reduced, and as the line profile becomes significantly larger than the instrumental width, the oscillations of the sinc profile are smeared out and disappear. The peak intensities decrease as the line gets narrower, because the areas under the lines are proportional to the original widths.

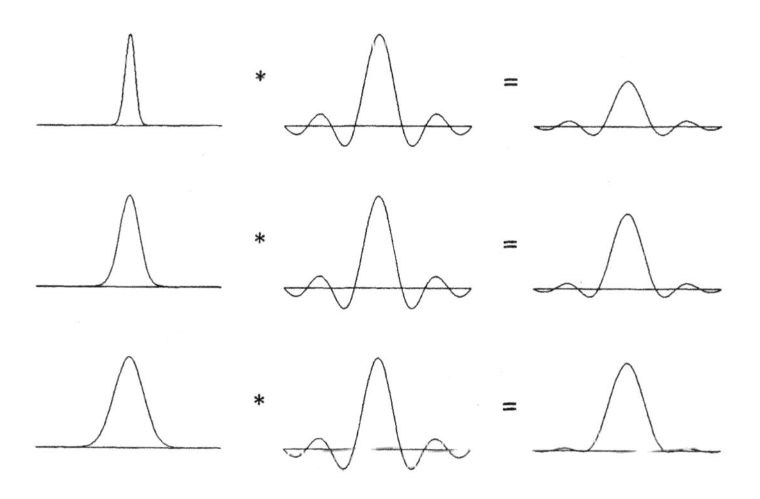

Fig. 5.4 The convolution of a fixed instrumental resolution width of 20 mK with gaussian line shapes of full widths 10, 20, and 40 mK.

It may be useful to look at the envelope of the interferogram and how it changes as the spectral line width changes in relation to the instrumental width. In a similar example with a fixed instrumental line width of 20 mK, but holding the area constant instead of the peak height, the interferogram envelope for a line width of 10 mK is truncated severely, that for the 20 mK width moderately truncated, and that of the 40 mK line truncated hardly at all, as shown in Fig. 5.5.

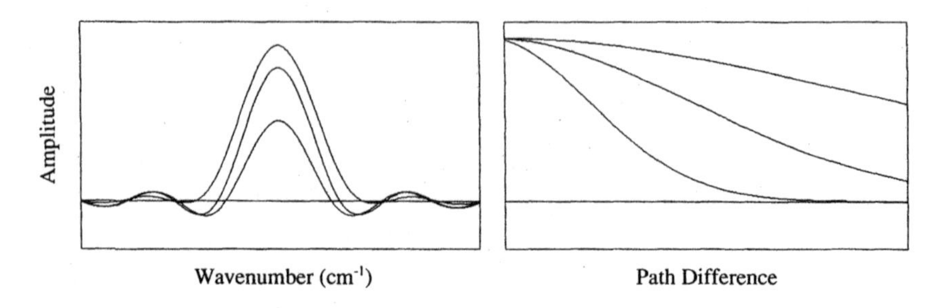

Fig. 5.5 Ringing in the spectral domain and truncation in the interferogram domain. Observed spectral lines. Corresponding interferogram envelopes. The uppermost spectral line profile corresponds to the lowest interferogram envelope.

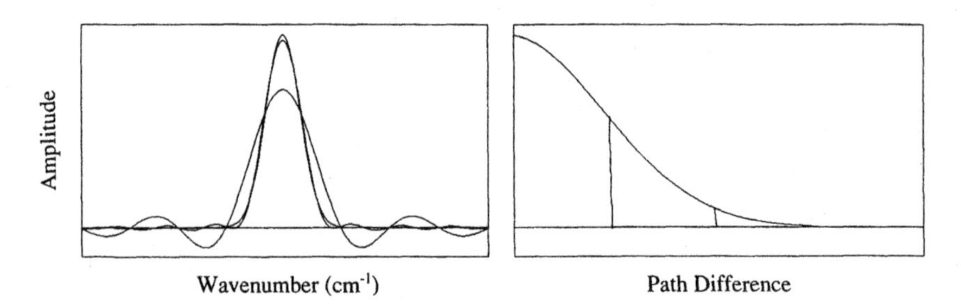

Fig. 5.6 A gaussian line of width 30 mK convoluted with sinc functions corresponding to resolution widths of 30, 20, and 10 mK, indicating that the necessary resolution width is about one-third of the observed line width. The sharpest line corresponds to the complete interferogram envelope. Vertical lines illustrate where the envelope is truncated to correspond with the broader lines.

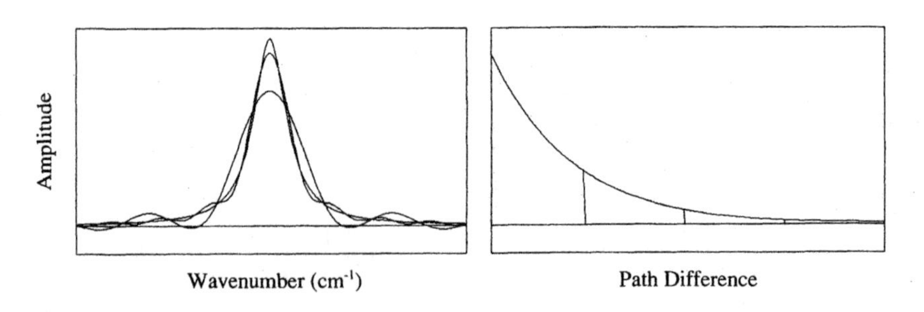

Fig. 5.7 A lorentzian line of width 50 mK convoluted with sinc functions corresponding to resolution widths of 50, 25, 16.7, and 12.5 mK. The nonzero magnitude of the transform at large path differences indicates that the necessary resolution width is roughly one-fifth of the observed line width. The sharpest line corresponds to the complete interferogram envelope. Vertical lines illustrate where the envelope is truncated to correspond with the broader lines.

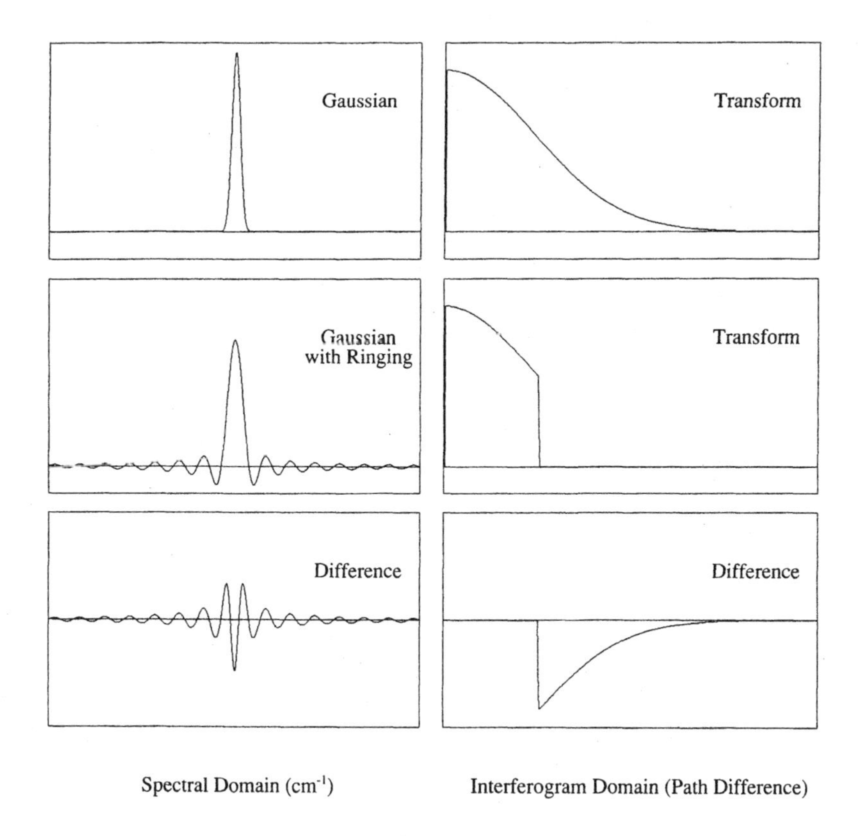

Fig. 5.8 Alternate method of interpreting ringing. Gaussian line and the envelope of its interferogram. Gaussian line with ringing and its truncated interferogram. Difference between the nonringing and ringing lines, and the difference of their corresponding interferograms.

Another interpretation of these instrumental effects is that the ringing results from the error introduced by the absence of the high-frequency components in the observed interferogram, as illustrated in Figs. 5.5 and 5.6. The conclusion to be drawn is that we aim for sufficient resolution to display a line without significant ringing. We can see how to do this empirically by keeping the line width fixed and varying the resolution width of the instrument. Figure 5.6 shows the corresponding lines and transform envelopes for a fixed line width of 30 mK and instrumental widths of 10, 20, and 30 mK. For the gaussian profile it is evident that a path difference corresponding to a resolution of one-third the line width ($1/2L = 1/3$ line width) gives an acceptable representation of the line profile. For a lorentzian shape one-fifth is required, as shown in Fig. 5.7. Most actual spectrum line shapes are voigtian profiles, a convolution of a gaussian and a lorentzian profile, but the results are qualitatively the same.

The *difference* between the ideal and the real transform, consisting of the high-frequency components as shown in Fig. 5.8, illustrates the origin of the line shape error due to path-difference truncation.

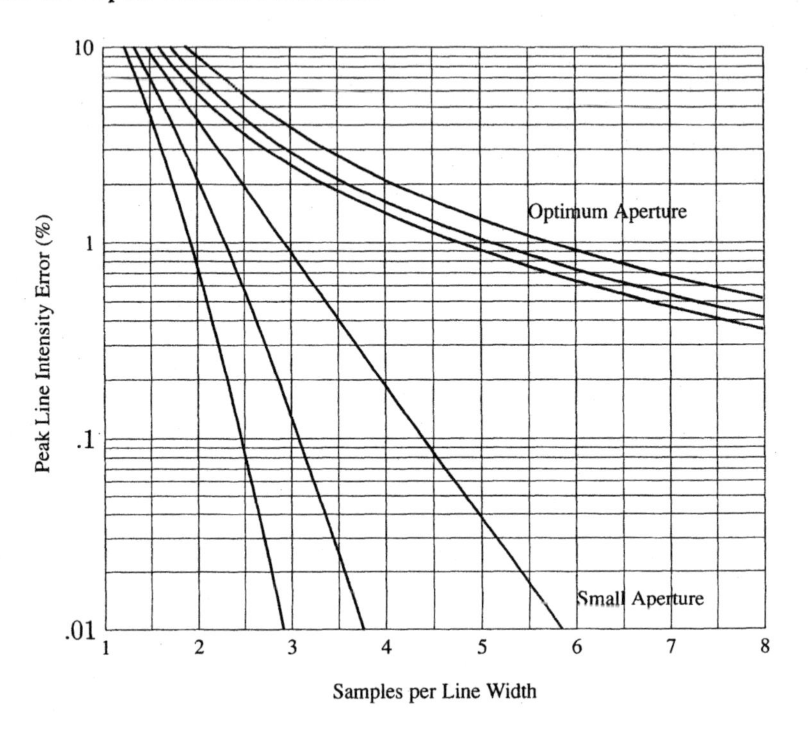

Fig. 5.9 Line peak intensity errors produced by the finite apparatus function. The three curves to the left, labeled Small Aperture, give the limiting error due to finite path difference when the finite aperture contribution is negligible; the upper three curves, labeled Optimum Aperture, are for equal contributions from the finite aperture and the finite path difference, with $k = 0.5$ assumed. In both cases, the left curve is for a pure gaussian and the right one for a pure lorentzian; the center curve is for equal gaussian and lorentzian components.

To summarize, the effects of the finite aperture and limited scanning distance depend on the aperture size, the shape of the line, and the ratio of the line width to the resolution width of the spectrometer. Figure 5.9 shows two families of curves, one for a negligible aperture size and differing profiles to show the effects of instrumental resolution alone, and the other for the maximum aperture permissible

($k = 1/2$). To get an idea of the percentage decreases in line peak intensities, note that for three samples (resolution elements) per line width and a negligible aperture size, the errors range from less than 0.01% for a gaussian shape to 0.9% for a lorentzian. Corresponding values for maximum aperture ($k = 1/2$) are 2.5% and 4%.

At a value of $k = 1/2$ the interferogram sees no more than half of the central lobe of the sinc function. To a very good approximation we can then replace it by a gaussian with similar curvature at the origin. A gaussian is much easier to work with mathematically because it has no discontinuities in slope, as does a rectangular function. In the spectral domain, this means that smearing with a narrow rectangle of width a is nearly the same as smearing with a gaussian whose FWHM is

$$a\sqrt{2 \ \ln \ 2/3} = 0.680a. \tag{5.23}$$

5.4 Nonuniform or Nonsymmetric Irradiation of the Aperture

So far we have been assuming that the entrance aperture of the FTS is circular, centered on the optical axis, and irradiated uniformly over its entire area. What if it is not? Is the effect of nonuniform or nonsymmetric irradiation going to make a difference in the shape of the interferogram and therefore modify the spectrum line shapes appreciably? There is no single easy answer, but we can discuss how to think about the problems.

For qualitative conclusions let us recall that there is a stretching factor in the wavenumber calibration owing to the size of the aperture, and that this must become a range of factors when the intensity distribution is not uniform over the aperture.

An arbitrary intensity distribution also changes or smears the k-value and distorts the interferogram, with the inevitable consequences of broadening and shifting spectral lines. It follows that a uniformly irradiated aperture is essential for accurate wavenumbers and line shapes of very sharp lines observed at the highest resolution.

To think about the problems in slightly more detail, consider an on-axis source whose image at the aperture is smaller than the aperture itself. Then the effective exit pupil — the image of the aperture at the detector — is the size of the source, not the size of the aperture. The effective k-value is smaller than the calculated one but there are no adverse effects on the line shape and the instrument resolution is greater than that calculated for the aperture.

In a different case, when the source image fills the aperture but the aperture is not centered on the optical axis of the FTS (an unlikely case in a well-designed instrument), Eqs. (5.9) to (5.12) must be considered. Equation (5.9) incorporates a weighting function that must be integrated over an off-axis circular aperture. A small circular part is on-axis, but the rest of the area is not. As a result the effective entrance pupil profile is asymmetric.

The effects of nonuniform and off-axis irradiation of the aperture have yet to be dealt with mathematically. In most practical cases the solution is to do the best possible with the sources at hand and to recognize that there may be some small distortions of spectral line shapes and equally small shifts in line positions. In our experimental work, even the most extreme variations in uniformity rarely produce shifts as large as 0.001 cm^{-1} in wavenumber. For very fast (f/3 to f/6) instruments of low resolution, a large aperture is acceptable and the problems and shifts may be significant.

The reader should note that all the examples of gaussian and lorentzian line shapes that we have used so far refer strictly to emission spectra. In absorption spectra it is the absorption coefficients that have such line shapes, and the spectral lineshapes are exponentials of these.

6

WORKING WITH DIGITAL INTERFEROGRAMS, FOURIER TRANSFORMS, AND SPECTRA

The precision required in calculating the Fourier transform that is at the heart of high-resolution FTS is far too great for any kind of analog technique. The development of the modern digital computer and such algorithms as the fast Fourier transform (FFT) are absolutely essential for high-resolution Fourier spectroscopy as we know it today. But *digital* invariably implies *discrete*, and so we must understand the constraints imposed by the operation of sampling the interferogram.

Most of our classroom experience is focused on the form and behavior of analytic continuous functions. Here, the measured data are a *sampling* of the theoretically smooth continuous analytic interferogram at uniform spatial intervals δx, as the mirrors traverse the path difference. Each datum point reflects only a small part of the basic interferogram. In effect, the continuous interferogram is multiplied by a sampling comb. Theoretically it is a comb of delta functions, but in practice the signal observation time is finite though small, and the mirrors are moving continuously.

There is a *sampling theorem* that specifies the conditions under which this discrete measurement of the signal is mathematically equivalent to the continuous interferogram and has a transform that is an accurate representation of the true spectrum, even though it, too, is digital, with the same number of total points as the interferogram. The theorem requires that the sampling frequency be at least twice the maximum frequency in the spectrum. Stated conversely, the spectrum must be limited to a maximum frequency, called the *Nyquist frequency*, equal to one-half the reciprocal of the finite sampling interval. The sampling frequency can be thought of as a sampling in time, in distance, or in wavenumbers.

However, we pay a price for this sampling of the interferogram. Its transform is not a single desired continuous analytical spectrum but is the spectrum convoluted with another comb of delta functions that have a large spacing in wavenumbers, given by the reciprocal of the interferogram sampling interval. The result is that the spectrum is not only sampled but also replicated, or *aliased*, many times over to infinity with this finite wavenumber spacing. The only condition that will permit us to recover a single spectrum from this series of replications is one that requires the spectrum to cover a finite range of frequencies. The range, including the negative frequencies, must be less than one-half the afore-mentioned spacing. When the spectral range is larger in extent than half this spacing, there is an overlapping of one part of the spectrum with another, somewhat like overlapping orders of a diffraction grating spectrum. A qualitative illustration is given in Fig. 6.1.

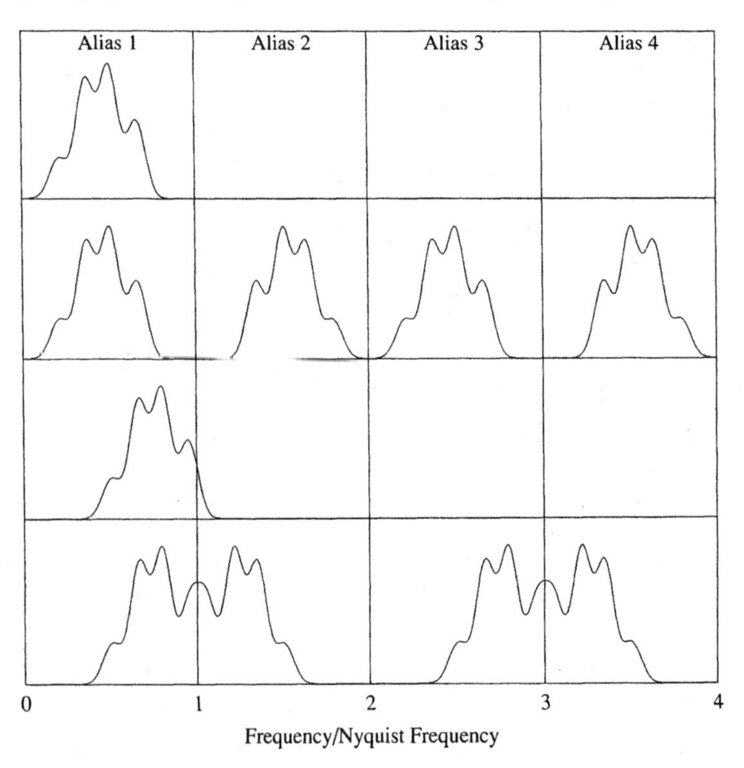

Fig. 6.1 Ideal spectrum from a continuous (unsampled) interferogram. Spectrum from a properly sampled interferogram, showing nonoverlapping aliases. The same spectrum envelope but moved to higher frequencies that exceed the Nyquist frequency. Spectrum from the improperly sampled interferogram.

Let's look in more detail at how all this comes about. Now is the time to refer again to Section 4.7.2, on the Dirac Comb, and the Section 4.9 examples.

6.1 Signals and Measurement

The sampled interferogram $I_s(x)$ is the product of the continuous interferogram $I(x)$ and the Dirac comb sampling function $III(x/\Delta x)$:

$$I_s(x) = I(x) \cdot III(\frac{x}{\Delta x}), \tag{6.1}$$

which contains only the values at the points $x = n\,\Delta x$. To take the transform of the sampled interferogram, we make use of the convolution theorem. The transform of a product is the convolution of the separate transforms, so the resulting discrete spectrum $S(\sigma)$ is a convolution of the desired spectral distribution and another Dirac comb

$$S(\sigma) = \widetilde{I}(\sigma) * III(\sigma\,\Delta x) = B_e(\sigma) * III(\sigma\,\Delta x). \tag{6.2}$$

The spectrum $B_e(\sigma)$is convoluted with a comb of delta functions having a spacing of $\Delta\sigma = 1/\Delta x$. In a convolution, when we pass a delta function through a spectrum we get the spectrum back again. When we have a comb of delta functions we get one spectrum for each delta function, spaced with the just-mentioned $\Delta\sigma$. In principle the spectrum is replicated indefinitely with this spacing.

For example, if the interferogram is sampled every 2.5×10^{-5} cm, then the maximum frequency (wavenumber) that can be recovered unambiguously is 20 000 cm^{-1}. Another way of putting this is to say that a sinusoid in the interferogram must be sampled twice in each wavelength in order to measure the frequency properly.

When the samples are too far apart and the nonoverlapping condition is *not* met, we may be in deep trouble from spectral aliasing. Figure 6.1c illustrates a spectrum that exceeds the Nyquist frequency and the aliasing that occurs (Fig. 6.1d). For a discrete line spectrum, aliasing can cause a true spectral line to be reproduced as a false line at a lower frequency, as shown in Fig. 6.2.

In the second line (Fig. 6.2) is a spectrum as it might look for a continuous analytic nonsampled interferogram, with both positive and negative frequencies a, b, c, and d. Above and below the true spectrum we show only the first of a series of positive and negative offset replicas, called *aliases*, which extend indefinitely in

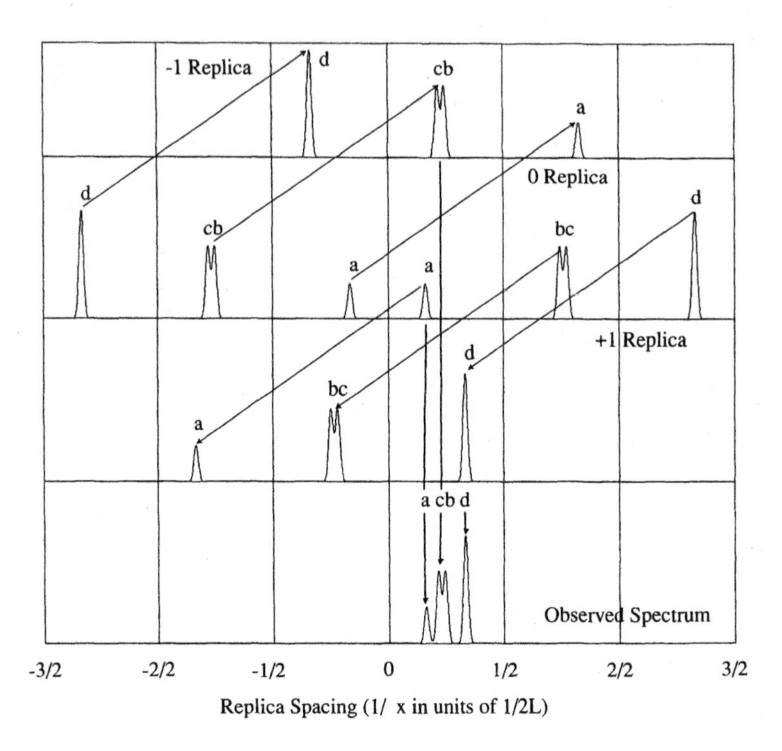

Fig. 6.2 Overlapping spectra from the first three replicas. The x-axis is given in units of the spatial sampling frequency, relative to the Nyquist frequency. The "Observed Spectrum" is replicated in all the other intervals as well. These replicas have been omitted from the figure.

both directions. Each replica is displaced by $1/\Delta x$, the reciprocal of the sampling frequency in wavenumbers. The effect is to superimpose the – 2/2 to – 1/2 region of the +1 replica onto the 0 to 1/2 region of the true spectrum, or, equivalently, to fold the 1/ 2 to 2/2 region of the true spectrum back on itself. Similarly the – 1 replica folds the +1/2 to +2/2 region of the true spectrum back onto its 0 to 1/2 region. The final observed spectrum is the sum of all these replicas over the interval 0 to 1/2. One way to think of this sum is to imagine the original spectrum plotted on z-fold paper, with the folds at $n/$ 2, and then folded up as shown in Fig. 6.3. We will call the interval from $(n-1)/2$ to $n/2$ the nth-order alias.

With this approach, when the spectrum is completely confined to any one of the folds, the spectrum is a true representation. In the example shown in Fig. 6.2, this is not the case. The sum of the true spectrum and the first replicas is plotted, and it is easy to imagine how complex the observed spectrum can become, and it

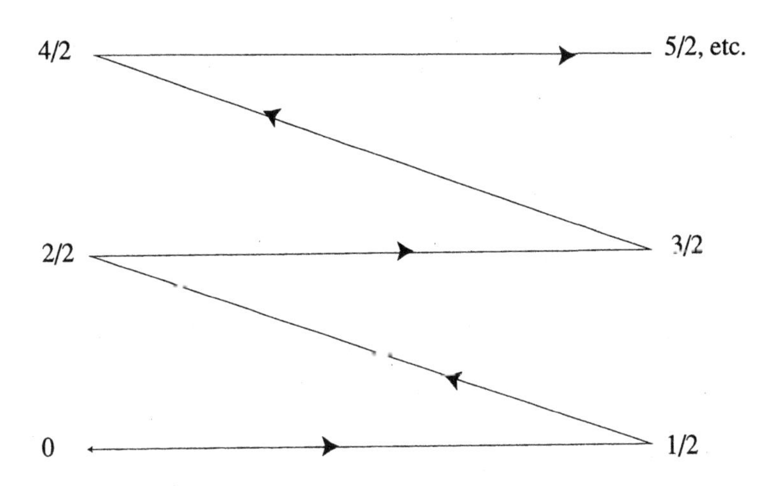

Fig. 6.3 Z-fold showing how spectral features in successive aliases can be superimposed.

may look completely unlike the true spectrum. Think of what the spectrum from one of Harrison's echelles looks like when the orders are not separated by a predisperser or other filter. It might be decipherable for an atomic spectrum with only a few lines, but certainly is not for a complex atomic or molecular spectrum.

To avoid aliasing in an extended spectrum, the spectrum often must be limited with an antialiasing low-pass filter that attenuates the undesirable frequencies. If the filtering is incomplete, the observed spectrum in the principal interval is the sum of all of the contributing aliases. But if the spectrum is relatively empty, it is often possible to interleave these replicas without any actual overlap of data and to reduce the number of samples required to represent the data. We have said that the antialiasing requirement is satisfied if the spectrum is completely confined to any one of the folds. This means that a nearly full spectrum (many lines filling a limited spectral region from beginning to end) must be observed in first order. But a spectrum that is limited to a narrow region far from zero, specified by $[(n-1)/n]\sigma_{\max} < \sigma < \sigma_{\max}$ can fit into the nth-order alias. Here, $\sigma_{\max}$ is one-half the Nyquist frequency. The importance of having a limited spectral range is that the number of samples of the interferogram required to determine the spectrum completely is reduced by n. In some cases this only reduces the time required to compute the final FFT, but in others it can be used to improve the resolution when the resolving power is limited by the number of data points that can be stored or to decrease the time needed to observe a single interferogram when the sampling rate is fixed or limited. It must be noted, however, that if a higher-order alias is used, it is essential that both the *optical* bandwidth for the spectrum and the *electronic* bandwidth for the noise be limited to that single order, since all aliases contribute to the final spectrum, including the noise.

6.2 Discrete Transforms

Now suppose that there are N consecutive sampled values in the interferogram, with a sampling interval δx. In the transform we can recover the spectrum only at N equally spaced points, the same number of points as in the interferogram. These points are spaced $2L/N$ apart, where $2L$ is the maximum path difference of the interferometer. This means that there are N nearly independent samples spaced one resolving limit apart, or we can say that there is one point for each *instrumental* linewidth $(1/2L)$. When we need to determine the spectrum between points, we can use interpolation methods to recover an approximation to the analytic spectrum. Later we will see how many samples per line width of the spectrum are required in order to fit the line profile adequately to determine wavenumbers, intensities, and damping.

The discrete Fourier transform (DFT) is a linear transformation that converts the N discrete measurements of the interferogram into N complex numbers through the expression

$$F(\sigma_k) = \frac{1}{N}\sum_{j=1}^{N} F(x_j)e^{-i2\pi x_j \sigma_k}, \tag{6.3}$$

where

$$x_j = (j-1)\delta x \quad \text{for } j = 1, 2, 3, \ldots$$
$$\sigma_k = (k-1)\delta\sigma \quad \text{for } k = 1, 2, 3, \ldots.$$

When we take N discrete frequency samples in the transform domain σ_k at intervals $\delta\sigma = 1/L = (N\,\delta x)^{-1}$ in the spectral interval $0 \to (N-1)/N\,\delta x$, then the kth frequency is

$$\sigma_k = \frac{k-1}{N\,\delta x}; \quad k = 1, 2, 3, \ldots, N.$$

The choice of frequency spacing is optimum, in the sense that the transform values are spaced one resolution width $(N\,\delta x)^{-1}$ apart.

The result, then, of a forward digital Fourier transform of N real data points is a set of N complex numbers, which contains a twofold redundancy in the computed set of transform values. The Fourier transform of real data has conjugate symmetry about both the origin and the Nyquist frequency such that

$$\widetilde{F}(\sigma) = \widetilde{F}(-\sigma)^*$$
$$\widetilde{F}(\sigma_N + \delta\sigma) = \widetilde{F}(\sigma_N - \delta\sigma)^*$$

and there is no new information in the data values above the Nyquist frequency. Figure 6.4 illustrates the symmetries of the real and imaginary amplitudes resulting from a Fourier transform of real data. More details and theorems are given in Chapter 4.

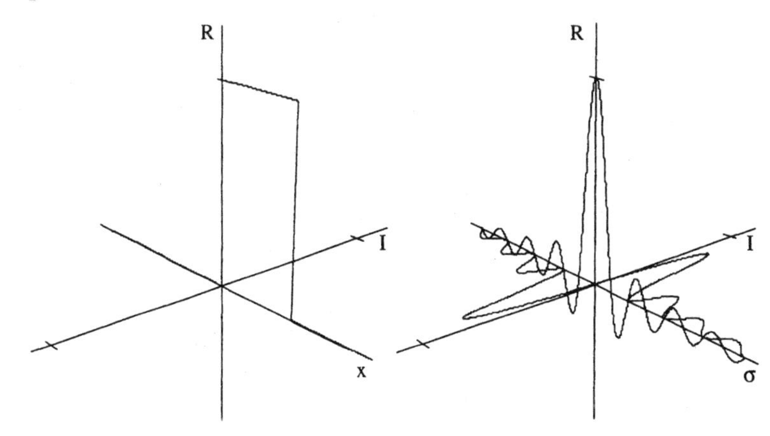

Fig. 6.4 The conjugate symmetry of the one-sided Fourier transform of real, asymmetric data illustrating the redundancy of information contained in the real and imaginary parts of the complex transform.

6.3 Interpolation and Filtering in the Spectrum

Now let's talk about some practical problems caused by having a sampled spectrum. The plot of a spectrum is a set of N points spaced $\delta\sigma = 1/2L$ apart and connected by straight-line segments. When there are only a few points for each spectral line width, the profile is segmented into a series of short straight lines giving an angular profile whose maximum may not be at the center of the line. Each point appears at a definite wavenumber, determined by the initial reference wavenumber and the dispersion. This linear wavenumber scale is quite adequate for many purposes, but often it is desirable to have the data sampled on a different scale with some other specified dispersion. We may want the spacing to be a multiple of some simple decimal fraction of a wavenumber, or we may want samples centered on a particular line (spectral lines seldom if ever have their peak intensities exactly on sampling points), or we may want the points equally spaced on a wavelength scale. All of these problems require that we interpolate values between the natural ones defined by $\delta\sigma = 1/2L$. Figure 6.5 illustrates some of the errors introduced when the interferogram is minimally sampled. The angular profiles arise from connecting the sampling points with straight lines. The smoother profiles result from having more sampling points that are much closer together.

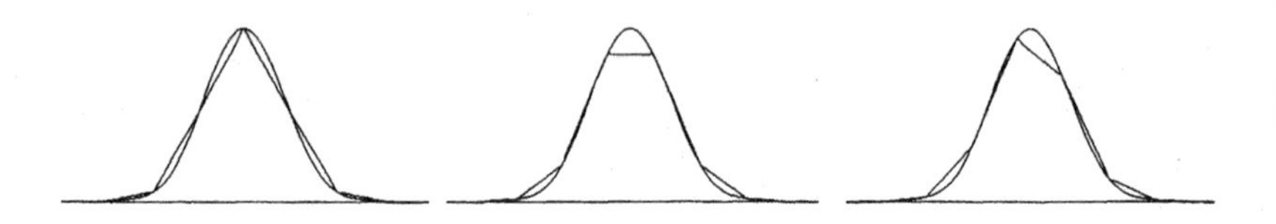

Fig. 6.5 Minimal sampling of a gaussian line profile: Two samples per FWHM centered on the peak value of the profile. The comparison profiles have eight samples per FWHM. Sampling grid shifted by 0.5 points and by 0.3 points. In the second and third cases, linear interpolation would produce an erroneous peak value, and in the third case an error in position as well.

The best way to start interpolation is to increase the number of points in the spectrum. We can do this by increasing the number of points in the *interferogram* before transforming it. The process is known as "zero-filling" or "zero padding." It is the extension of the interferogram by adding points with values of zero. Typically the extension is by a factor of 2 longer than is essential for data acquisition. As a side issue, it should be noted that zero-filling is often used for a different purpose. The common form of the FFT requires that the total number of samples be a power of 2, so the usual practice is to extend the interferogram with zeroes to fulfill this condition.

The consequences of the additional zeroes are to change the number of points in the interferogram, to increase its length, and to change the dispersion in the spectrum by increasing the number of points. The length of the spectrum is unchanged. Put another way, zero-filling in the interferogram interpolates the spectrum by increasing the number of sampling points.

What we would really like, of course, is to have an exact curve with an infinite number of points drawn through the N experimental points, but extending the number of zero-filled points to infinity is not a practical solution. In any event, it is always true that $\sigma_{\max} = 1/(2\,\delta x)$, so the extended transform gives points with a spacing $\delta\sigma = 2\sigma_{\max}/N = 1/N\,\delta x$ with $N = 2^m$. The comparison profiles in Fig. 6.5 illustrate the effectiveness of zero-filling the interferogram.

Next consider some interpolation schemes that are known to be very practical and common. Qualitatively, to see how good they are we first settle on a method of interpolating in spectral space (what sort of curve to pass through the points). Then we take the transform of this curve and see how closely it compares with the ideal rect function in interferogram space, as discussed earlier. The more closely it approaches a rect function, the better the interpolation, and conversely.

The simplest of these curves are the polynomial interpolation formulas that result from finding the polynomial of order n, that passes exactly through $n + 1$ adjacent points and using this polynomial to give the desired values over the interval between the two points closest to the center of the $n + 1$ region. The effect of all these interpolation schemes or formulas can be described exactly in terms of a function that is to be *convoluted* with the spectral data points and whose *transform* therefore defines the effective filter being applied to the interferogram (multiplies the interferogram).

As an example, consider the simplest case of all, linear interpolation between adjacent points in the spectrum

$$f(i, z) = f_i + z(f_{i+1} - f_i) = (1 - z)f_i + zf_{i+1}, \tag{6.4}$$

where z is the fractional distance between points i and $i + 1$. The latter form provides the shape of the equivalent convoluting function, a triangle of unit height and a base two units wide, as shown in the first line of Fig. 6.6.

This is *not* a good approximation to the *sinc* function, even though it may look reasonably close, because its transform is a far cry from the desired rectangular function. If we recognize this triangle as the self-convolution of the rectangular Π function, we can immediately write down its transform: $\text{sinc}^2(x)$, which is a very poor approximation to the desired rectangular sharp cutoff filter. The interferogram is multiplied by a $\text{sinc}^2(x)$ function instead of a rectangular function! Obviously, linear interpolation between points in the spectrum seriously compromises the accuracy of minimally sampled data, and some curious artifacts can be produced by its careless use, especially with underresolved (ringing) spectra. To be more quantitative, the maximum intensity error in interpolation comes at the center of a line that falls halfway between two sampling points. Linear interpolation then gives the center the same value as the average height at those two points, as illustrated in Fig. 6.5. In the case of a gaussian line with a FWHM of W, we can write

$$e^{-(x/a)^2} = e^{-4 \ln 2x^2/W^2}, \tag{6.5}$$

so if we have three sampling points per FWHM, a typical number, then we can have $x = \pm W/4$, and

$$e^{-(\ln 2)/4} \sim 1 - \frac{\ln 2}{4}. \tag{6.6}$$

The error from linear interpolation can be greater than 16% of the line strength. For a lorentzian-shaped line, the result is similar, except that ln(2) is replaced by 1 in Eq. 6.6, so the error reaches 22%. The lesson is clear: *Avoid linear interpolation of FTS spectra.*

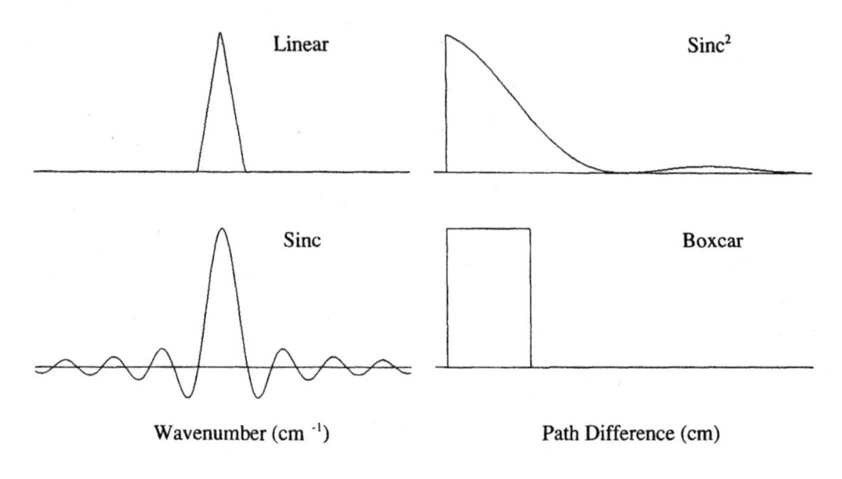

Fig. 6.6 Linear interpolation function and corresponding filter function (multiplier of the interferogram). Exact interpolation function (sinc) and filter function.

With today's computers it is easy to consider more sophisticated techniques, so there is *no longer any excuse to use linear interpolation.* To begin with, we can use higher-order polynomials. Since it is most convenient to use an interpolation formula over an interval *between* points rather than an interval one point wide *centered* on a point, we use odd functions. Also, even-order polynomials should be avoided because the equivalent convoluting functions are asymmetrical in interferogram space. The filter responses of the first three odd polynomials are given by the expressions

Linear (2 points) $\operatorname{sinc}^2 x$

Cubic (4 points) $\operatorname{sinc}^4 x - \frac{2}{3}\operatorname{sinc}^2 2x + \frac{2}{3}\operatorname{sinc}^2 x$

Quintic (6 points) $\operatorname{sinc}^6 x + \operatorname{sinc}^4 x - \operatorname{sinc}^2 x \cdot \operatorname{sinc}^2 2x + \frac{3}{10}\operatorname{sinc}^2 3x$

$$-\frac{4}{5}\operatorname{sinc}^2 2x + \frac{1}{2}\operatorname{sinc}^2 x.$$

Maximum errors produced by interpolation of gaussians of different widths by linear and quintic formulas are shown in Fig. 6.7. The error for a polynomial of order n falls off as $1/W^{(n+1)}$. Figure 6.8 illustrates several interpolation functions.

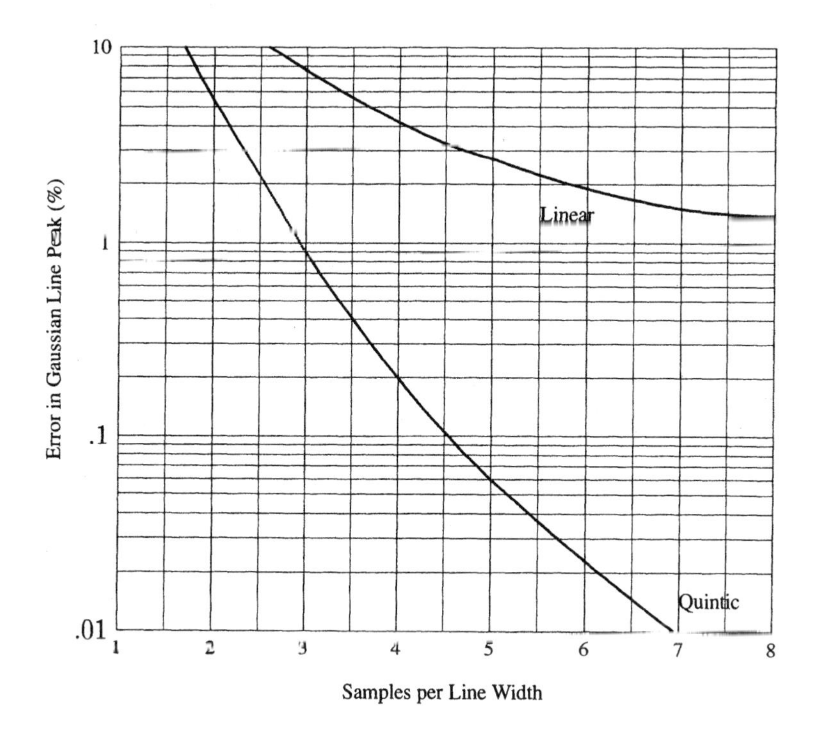

Fig. 6.7 Peak intensity errors produced by polynomial interpolation.

Many spectra have signal-to-noise ratios of 1000:1 or better, so our usual procedure is to do the following:

a. Observe with enough resolution to provide three points per FWHM for a typical line.

b. Extend the interferogram a factor of 2 with zeroes so the final spectrum has six points per FWHM.

c. Use quintic interpolation routines built into the standard software.

A good alternative is to return to the ideal sinc function and modify it in such a way as to make it practical without destroying its accuracy. The first step is to apodize it — cut off its feet — by multiplying it with a suitable function to depress

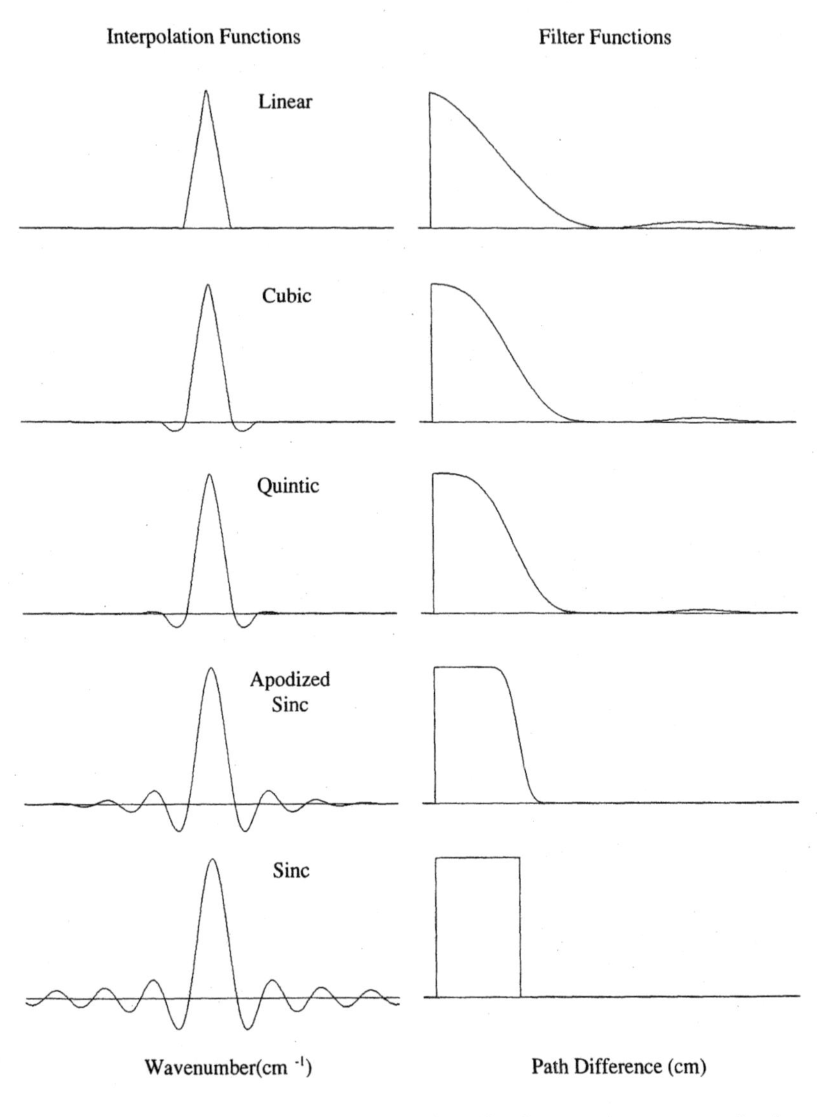

Fig. 6.8 Properties of several interpolating functions: linear (2 points), a rather poor approximation of a sinc function and its transform; cubic (4 points; quintic (6 points); gaussian apodized sinc (21 points); sinc, the exact interpolation function (Fourier interpolation).

the far wings. A gaussian is nearly ideal for this. Then when the function has fallen to a sufficiently low value, it can be truncated abruptly without causing significant error. In some of our earlier work, we often used

$$\text{sinc}\,\sigma \cdot e^{-0.06\xi^2}, \qquad -10 \leq \xi \leq 10, \tag{6.7}$$

which requires a 21-point convolution.

The filter response and convoluting function for all of the interpolation schemes just discussed are illustrated in Fig. 6.8. They show the progressive convergence of the polynomial functions toward the ideal sinc function. As illustrated in Fig. 6.8, the accuracy of the interpolation process is largely proportional to the number of terms in the approximation, with linear interpolation producing the worst result, Fourier interpolation produces the best result, and the other functions producing intermediate results. A detailed examination of the convoluting functions illustrates the addition of sidelobes as higher-order polynomials are introduced into the interpolation function. Similarly, in the transform domain, each additional term increases the slope of the filter truncation and increases high-frequency harmonics.

6.4 Apodization

Apodization is a concept and a process describing a modification of the interferogram envelope in order to produce "nice-looking" spectral lines relatively free from noise. Literally it means "cutting the feet off." Let's see why we need to be concerned about this.

An infinitely long interferogram reproduces the spectrum exactly. When the incoming radiation is truly monochromatic, the spectral lines are delta functions, and the transform of an infinite interferogram verifies this fact. It takes infinite resolution to reproduce delta functions. When the interferogram is terminated at some finite length, then monochromatic spectrum lines have finite widths and are accompanied by a series of sidelobes, called *ringing*. In practice, quasi-monochromatic lines also have sidelobes if the instrumental width is less than that of the spectral lines, as mentioned in Chapter 5. We explain this by noting that a finite length of interferogram implies that we have multiplied it by a rectangular function. In the spectral regime this is equivalent to convoluting the true spectrum with a sinc function. When the spectral lines are sharper than the instrumental function, the sinc function sidelobes are easily visible. When the lines are broader, the sidelobes are less noticeable and may not significantly affect the spectrum appearance. The questions are: Can we modify the interferogram to reduce or eliminate the sidelobes? Can we multiply it by a smoothing or apodizing function, to achieve the desired good-looking spectral lines without seriously modifying their intrinsic shapes?

A way of thinking about the process of apodization is to consider spectral line widths and their influence on interferogram modulation. Look at a spectral line with a finite width and gaussian shape. Its wavetrain has a gaussian-shaped envelope that is limited in length, for practical purposes. (Remember that noise is everywhere and that when the envelope amplitude reaches the noise level, nothing is gained by extending the wavetrain further.) The interferogram of this line therefore has a gaussian-shaped envelope that tends to zero near a path difference that makes the instrumental line width equal to the spectral line widths. The interferogram is said to be *self-apodized*. There is no problem here. Look at Fig. 6.9.

Now look at Fig. 6.10, which shows the same spectral lines from a much hotter source with broader spectral lines. The interferogram is again self-apodized and is effectively zero at a much shorter path difference.

To reiterate and illustrate what we have just said, if it is not possible to extend the interferogram to make the instrumental width less than the line width, we are in trouble. Every line shows sidelobes of the sinc function and is said to show "ringing," illustrated in Fig. 6.11. The ringing dies off only as $1/\Delta\sigma$, and this greatly increases the problems of blending in a complex spectrum because both strengths and positions of neighboring lines are affected. To the uninitiated, the sidelobes may appear to be real lines, especially if the spectrum is a bit noisy.

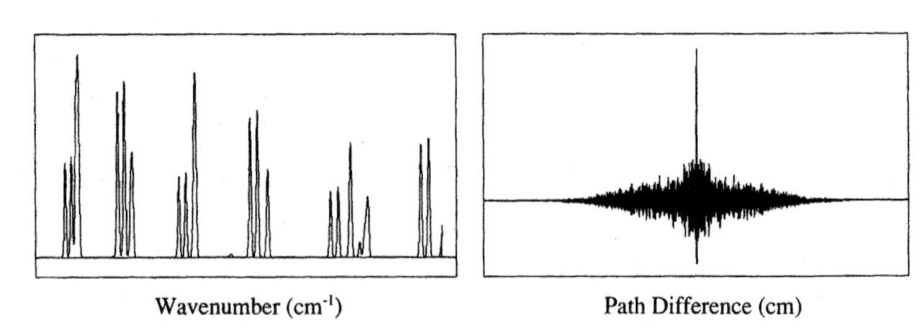

Fig. 6.9 Spectrum of the nitrogen molecular ion near 23 400 cm^{-1}. The accompanying interferogram. Note the self-apodization of the interferogram.

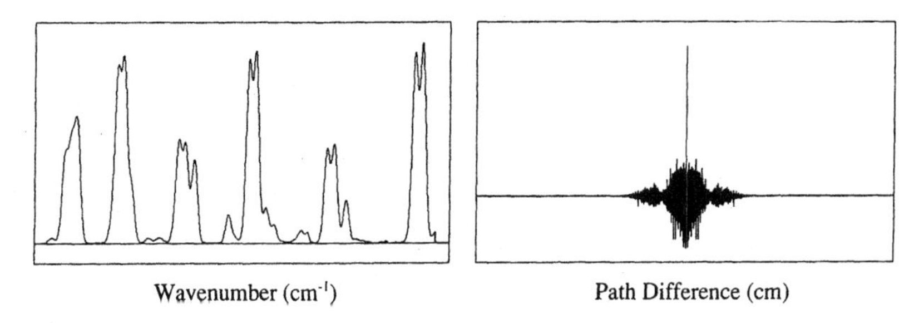

Fig. 6.10 Spectrum with broader lines and consequently a shorter self-apodized interferogram. Interferogram.

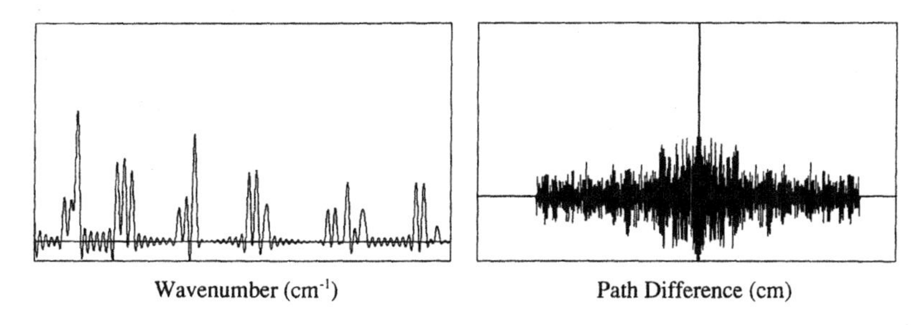

Fig. 6.11 Spectrum showing ringing and reduced resolution, owing to a sharp cutoff of the interferogram before it reaches zero amplitude. Compare with Fig. 6.9.

Given an appropriate warning, the user can discount the appearance (as software people put it, "*It's a feature, not a bug.*") But in principle the most effective way to handle the problem is to increase the resolution until the ringing is lost in the noise.

The curves in Figs. 5.6 and 5.7 can be used to estimate the required resolution. As a rule of thumb we use three points/FWHM for lines that are predominantly gaussian and four or five for lines that are strongly lorentzian. Unfortunately, in practice the required resolution may be outside the capabilities of the instrument at our disposal. Also, resolution is expensive in both time and resources. In some cases, however, the desired information is available from spectra of lower resolution if they are handled properly. We shall discuss this further under observing strategies.

If we cannot increase the resolution but need to eliminate the ringing, then the next most common solution is to *apodize* the spectrum by multiplying the interferogram with a function that smoothly reduces its amplitude to a small value or zero before the end is reached so that the discontinuity there, which is the major source of the ringing, is minimized. As a general rule, any discontinuity in the value of the envelope leads to an apparatus function that falls off asymptotically as $1/\Delta\sigma$.

The price paid for this improvement is of course some degradation in resolution. The effective extent of the interferogram has been reduced, so the resulting lines are broader. This broadening is especially onerous in the case of some of the simpler functions that have been used for apodizing, such as the triangular function. Considerable effort has historically been put into the development of apodizing functions that minimize the ringing while reducing the resolution as little as possible. A comprehensive discussion of such functions can be found in many references, but unwary users often overapodize because they forget that the line shapes discussed are idealized and do not include the effects of self-apodization which is always present. It should also be noted that our old friend the gaussian is not too far from optimum as an apodizing function and has the great advantage that its effect on any line that can be well represented by a voigtian function is easily corrected for. We reiterate that with today's computers there is no longer any excuse to use a linear (triangular or trapezoidal) apodizing function and to tolerate the intensity and line shape distortions that it introduces. The spectral lines resulting from using various apodizing functions on the truncated interferogram are shown in Fig. 6.12.

In the final analysis, however, the decision about whether or not to apodize depends to a large extent on the end use of the data. In atlases and other visual displays intended for inexperienced users, it is important to suppress any visible ringing. When finding line positions, the application of simple peak-finding algorithms to relatively complex spectra usually benefits from some lighter apodization so that only the strongest lines are accompanied by a few sidelobes. A heavily apodized spectrum as might be prepared for an atlas should not be used for line finding. We should always fit the data with a model, even if we only need line positions. With some models, and a well-chosen method of fitting with a nonlinear least-square algorithm, the effect of the known finite apparatus function can be included in the model in a natural way and there may be no need for apodization. The information in the unapodized interferogram is utilized, and the interaction between adjacent lines is accounted for automatically. A detailed discussion of this procedure is given in Chapter 9. Nevertheless, we cannot stress too strongly that *any* apodization modifies the spectrum, reduces the resolution, and otherwise degrades the intensity, width, shape, and area of all spectrum lines, often in unsuspected ways.

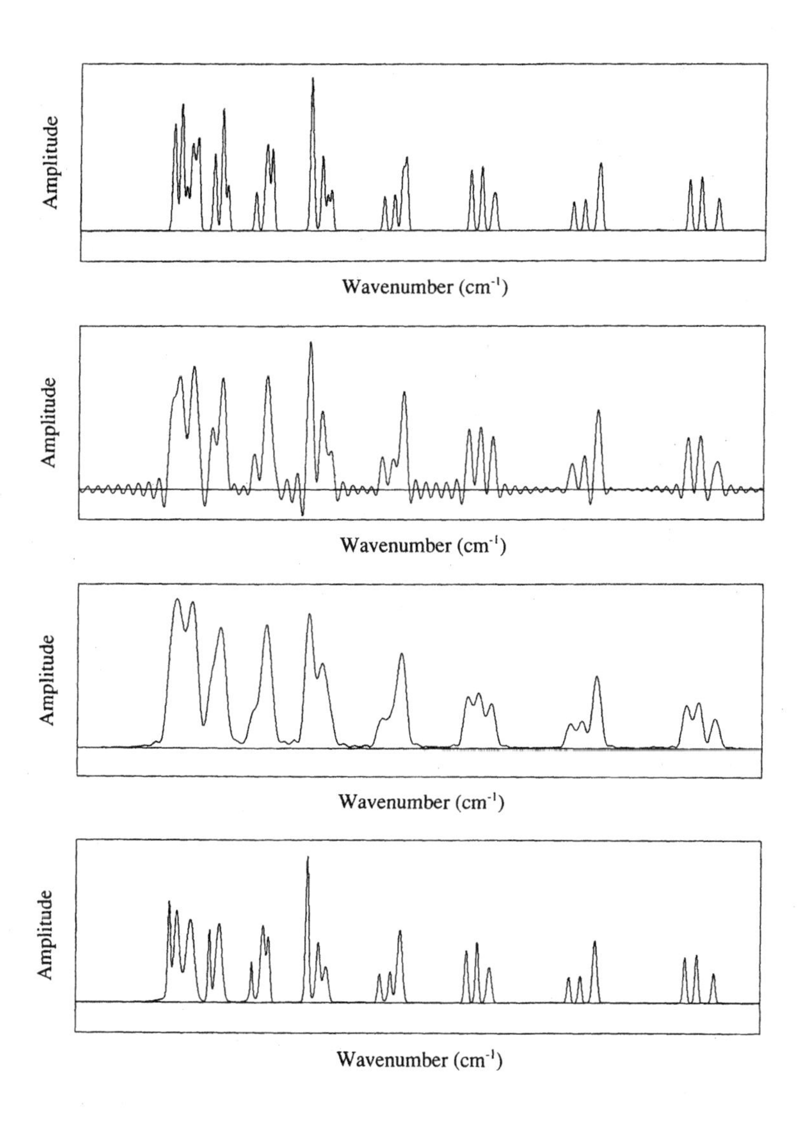

Fig. 6.12 Fitted spectrum line profiles resulting from different interferogram apodizing functions. True spectrum from interferogram with self-apodization only. Spectrum from truncated interferogram with no apodization. Spectrum from truncated interferogram with triangular apodizing function. The ringing is gone, but so is the resolution. Spectrum from truncated interferogram with gaussian apodizing function and filtered fitting of the spectrum. The ringing has been eliminated and the resolution retained.

6.5 Preparation for Making Transforms

Fourier analysis of actual data sets involves more work than simply applying a numerical algorithm to a data set. Several practical features of numerical data processing require our attention.

6.5.1 Common Artifacts

Occasionally, periodic time variations in the source produce long-period oscillations in the interferogram and low-frequency spikes in the transform that appear as sidebands on strong lines. Pay attention to all places where such effects might enter the data. For example, 60-Hz oscillations from power lines can easily enter into data, although they are harmless unless the artifacts they produce in the spectrum fall on top of some other spectral feature. Mechanical oscillations such as the rotation of the plasma in an inductively coupled plasma light source can also lead to spurious spectral features.

6.5.2 Zero Mean Interferograms, End Effects, and Masking

To avoid creating low-frequency distortion, it is useful to subtract the mean value from a data set before transforming. This removes the mean value discontinuity at the origin of the transform. Of course the ends of the data set must be treated with care to avoid distorting the transform unnecessarily.

The discontinuities at the ends of a data set present a problem unique to the FTS. The rapid change in the interferogram is incompatible with a measurement that has a finite bandwidth. The effect of the discontinuities is to introduce oscillations (Gibbs phenomena) when the data are interpolated or filtered. In practice, the edge discontinuities are suppressed by imposing a smooth transition from the mean of the measured values to zero, through multiplying the data by a set of weights that is termed a *data window* or *mask*. A typical data window masks the end regions by multiplying the data by a cosine bell curve that tapers the first and last 10% of the data to zero. Masking should be applied only to zero-mean data, for without this restriction the end discontinuities will be larger than necessary. Masking the data produces some smoothing in the transform domain and reduces the overlap between the replicas of the spectrum.

6.6 Procedure for Reliable Transforms

The material in the preceding sections can be summarized by the following sequence of operations that produces reliable transforms and avoids many of the pitfalls that arise from blindly applying a numerical Fourier transform to a data set.

1. Sample the data at a rate such that the signal is oversampled, to prevent aliasing of the replicas of the basic transform,

$$\frac{1}{\delta x} \geq 2\sigma_{\mathrm{max}}$$

2. Include a real-time filter with a cutoff frequency no greater than one-half of the sampling rate in the sampling instrument, to avoid aliasing by noise.

3. Subtract the mean value from the data before transforming, to eliminate the low-frequency spike.

4. If you must smooth the ends of the interferogram by apodization, do not use a triangle or trapezoid as the apodizing function. The gaussian is far more suitable.

5. Extend the data by zero-filling, to increase the spacing between the replicas of the data segments and to ensure that the sampling theorem is not violated when taking derivatives during the processes of interpolating or fitting the spectrum.

7. Transform the data.

8. Whenever there is any doubt about how the preceding procedures affect or modify the spectrum, construct a sample spectrum of lines whose properties are all well defined, form an interferogram, process it just as you expect to process the "unknown" interferogram, and see how accurately the processed spectrum matches the original sample. This is another way of saying, construct an instrumental function, and see if it meets with your standards of spectral analysis.

7

PHASE CORRECTIONS AND THEIR SIGNIFICANCE

7.1 The What and Why of Phase

In earlier chapters we worked with ideal interferograms, showed how real ones differ from the ideal and the consequences of the differences, and discussed modifications made necessary by digital sampling. There is yet a further set of real-world problems in the interferograms and in their use. We know that the true spectrum is real and positive, and we assume that it can be represented by a symmetric function and by a cosine transform of a symmetric interferogram. The sine transform representing the complex part is zero. The phase is zero. But in fact the measured interferogram is not a real symmetric function, because experimental, instrumental, and computational limitations introduce asymmetries into it. Complete reconstruction of the spectrum requires a complex Fourier transform, from which the true spectrum must be recovered. If only the real part of an uncorrected complex spectrum (phase not equal to zero) is used as the representation of the true spectrum there can be serious errors in line shapes, sizes, positions, and signal-to-noise ratios. Therefore, to get spectra from the FTS that are limited in quality only by source or detector noise, the phase must be determined and corrected with a precision that greatly exceeds its day-to-day reproducibility. These corrections must be done separately for each interferogram or coadded set of interferograms and must be deduced from the data contained within the interferograms.

To put the problem into perspective, we note that with a high- resolution FTS we can determine emission line wavenumbers with an accuracy of a few parts per billion, limited by the signal-to-noise ratio of the line and the availability of suitable reference lines. In many cases the phase must be determined with an error, in radians, that is approximately the reciprocal of the signal-to-noise ratio in the computed spectrum. With a ratio of 1000, the phase must be corrected to the order of 1/1000 radian.

We might ask, "Why not take the modulus of the complex transform rather than correct the phase, since the process should give exactly the same spectrum with a simpler computation?" The answer lies in the fact that the signal-to-noise ratio of the spectrum is then degraded, because the imaginary part of the transform contains as much noise as the real part. Using the modulus carries over this part of the noise into the "real" spectrum, unlike the process of phase correction.

7.2 Origins of Asymmetries in the Interferogram

Asymmetries in the interferogram arise from a combination of causes. A prime one is the frequent selection of a one-sided interferogram rather than a complete two-sided (equal-sided) interferogram, but there are other causes that produce significant asymmetries. No source has a truly stationary signal intensity over the time required for recording even a single interferogram. The center of the zero-path fringe of a symmetrical interferogram usually does not coincide exactly with one of the points at which the interferogram is sampled. The apodizing function is often not symmetrical about the zero optical path difference. The optical path difference may be frequency dependent, owing to dispersive effects, so there is no single well-defined position of zero path difference for all frequencies. Such dispersive effects may be optical (small differences in the beamsplitter and compensator thickness) or electrical (frequency- dependent delays in the amplifying and filtering electronics). When dispersion is present, the interferogram is not symmetric about the center of the largest fringe. All these asymmetries result in what are called *phase errors*.

7.3 From Asymmetries to Phase Errors

We begin with a real spectrum of finite extent $B(\sigma)$. Given such a spectrum as input, an ideal FTS produces a sampled interferogram described by

$$I(x_n) = \sum_{j=1}^{N} B(\sigma_j) \cos (2\pi\sigma_j x_n), \tag{7.1}$$

where $I(x_n)$ is one of the samples of the interferogram, taken when the optical path difference is x_n. For mathematical convenience we reflect the spectrum about

$\sigma = 0$, as discussed earlier (Fig. 3.5) and write the interferogram in the more general form

$$I(x_n) = \sum_{j=-N}^{N} B(\sigma_j) e^{i2\pi\sigma_j x_n}, \tag{7.2}$$

which is more convenient when we deal with phase shifts.

Now, what exactly *is* a phase shift, and how is it produced? The presence of a phase shift implies merely that the exponent in Eq. (7.2) does not go to zero when x is zero and that ϕ has a wavenumber dependence. We rewrite the equation as

$$I(x_n) = \sum_{j=-N}^{N} B(\sigma_j) e^{i(2\pi\sigma_j x_n + \phi_j)} \tag{7.3}$$

where ϕ_j is the phase shift corresponding to the frequency σ_j. In most cases, the phase shifts come from one of the following sources:

1. Our sampling grid has no point that coincides with $x = 0$. Then $x_n = n\,\delta x + \alpha$, and the exponent is

$$2\pi\sigma_j(n\,\delta x + \alpha) = 2\pi\sigma_j n\,\delta x + 2\pi\sigma_j\alpha, \tag{7.4}$$

or $\phi_j = 2\pi\sigma_j\alpha$. This is the very common case of a phase shift proportional to the optical frequency.

2. There is some unbalanced *dispersion* in one arm of the interferometer. For example, the compensator plate may not have quite the same thickness as the beamsplitter. In this case, the exponent becomes

$$+2\pi\sigma_j(x_n + r_j d), \tag{7.5}$$

where r_j is the index of refraction of the unbalanced material at frequency σ_j and d is its thickness. If r_j were constant, this would fall under the previous case. The new element here is the *dispersion* of r_j. If the observation covers a fairly wide spectral range or is close to an absorption edge of the material, we can easily have $\delta r/r$ as large as 2%.

3. If the interferogram is really sampled in *time* and connected to the desired spatial positions by a carriage speed, then electronic circuits may produce time delays that vary with frequency.

A specific example may help to put the phase problem in perspective. Figure 7.1 shows the central 512 points of an extended interferogram taken on the Kitt Peak instrument. The spectrum is the infrared emission spectrum of an OH diffusion flame between 1800 and 9000 cm^{-1}. Low-resolution amplitude and phase spectra obtained from the equal-sided interferogram are shown in Fig. 7.2.

In the regions where the signal-to-noise ratio is appreciable in the amplitude spectrum, the phase spectrum is well defined and approximately horizontal. The curvature indicates that the location of the central fringe shifts slightly with frequency, which is characteristic of phase errors resulting from unbalanced dispersion. When the amplitude is very small, the phase fluctuates wildly because there is insufficient information to determine it unequivocally.

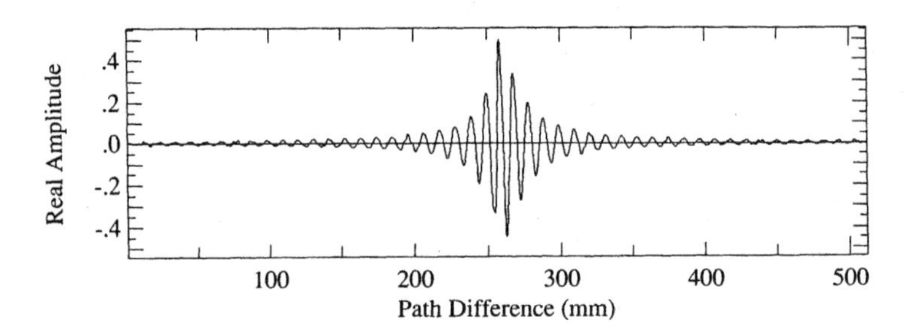

Fig. 7.1 Central 512 points of an interferogram.

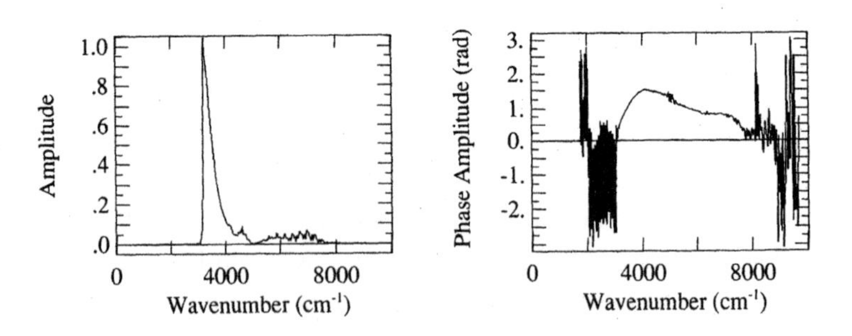

Fig. 7.2 Low-resolution amplitude and phase spectra from an equal-sided symmetric interferogram with a sampling point at the origin of the central fringe.

To illustrate the sensitivity of the relationship between the symmetry of the interferogram and the spectrum that results, the interferogram in Fig. 7.1 can be shifted. If in the sampling of the interferogram a sample is not taken at the true

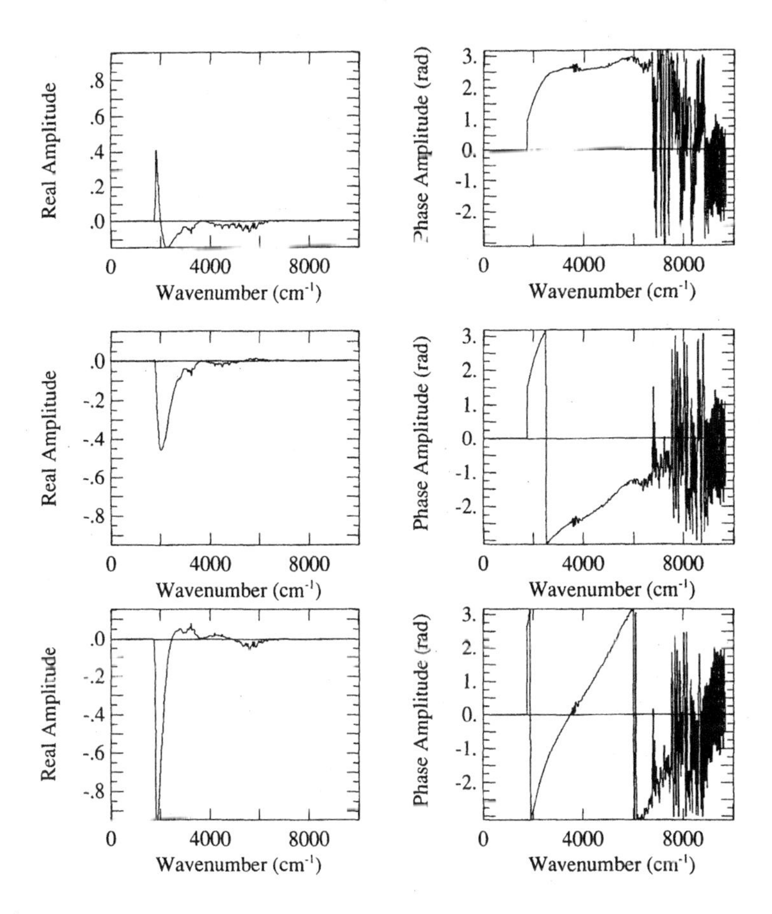

Fig. 7.3 Low resolution amplitude and phase spectra showing the effects of shifting the origin of an equal-sided symmetric interferogram: Origin shifted by 1 point (top). Origin shifted by 2 points (middle). Origin shifted by 4 points (bottom).

origin, the position of the white-light fringe, and it is not corrected for, then serious difficulties arise when the interferogram is transformed. In the case where the central fringe is displaced by one or several points from the absolute position, as can occur easily with unequal-sided interferograms, the shift introduces a rotation into the resulting spectrum. In Fig. 7.3 are shown the effects of shifting the origin

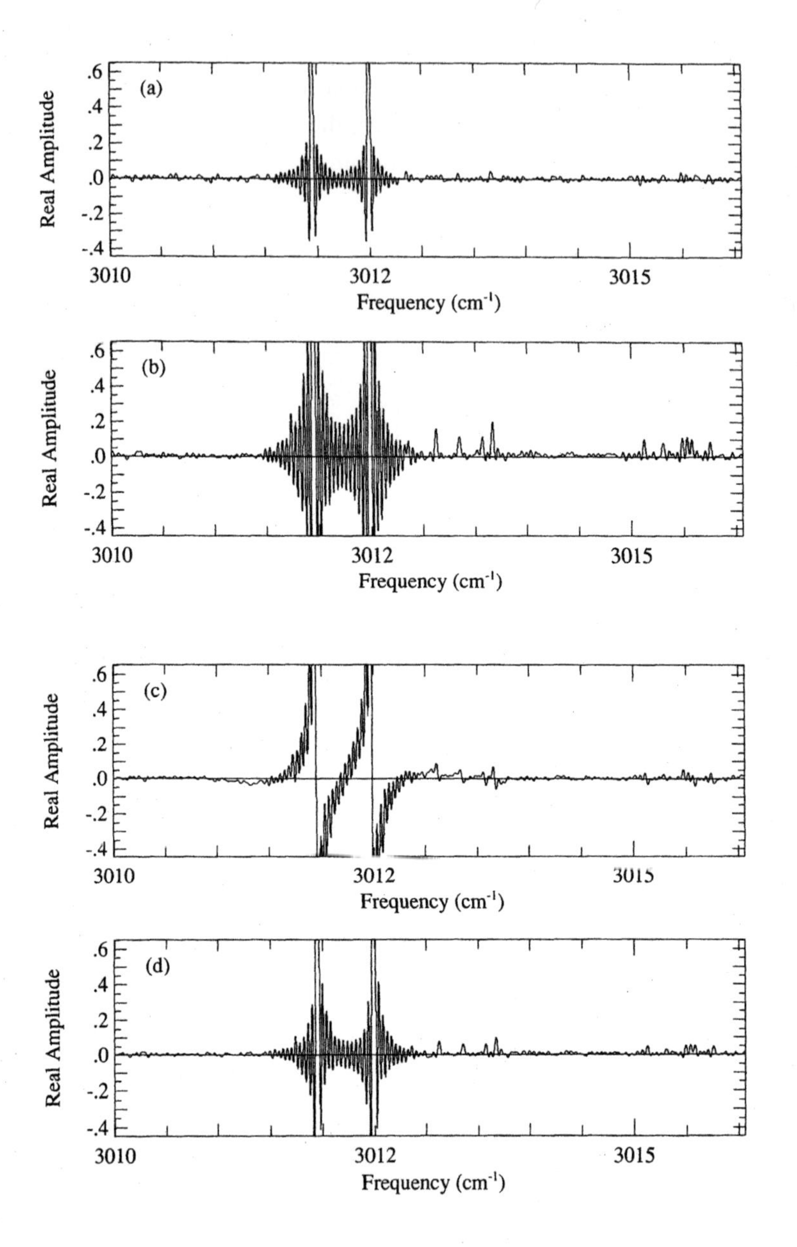

Fig. 7.4 Real spectrum resulting from a cosine Fourier transform of an equal-sided interferogram. (a) Transform without phase correction. (b) Transform with phase correction. (c) Real spectrum resulting from a cosine Fourier transform of an unequal-sided interferogram, without phase correction. (d) Transform of an unequal-sided interferogram with phase correction. The decrease in amplitude between the phase corrected spectra is real and reflects a change in integration time or number of points by a factor of 2.

by 1, 2, and 4 points. A shift by only one point introduces a rotation of π in the phase spectrum as illustrated by the vertical discontinuities in Fig. 7.3.

A high-resolution spectrum illustrating dramatic improvement as a result of phase correction is shown in Fig. 7.4. The intensity with phase correction is almost twice that without correction, because the phase angle happens to be nearly 45 degrees. Many weaker lines, unobserved in the real part of the spectrum without phase correction, now stand out distinctly. The transform of the unequal-sided interferogram illustrates the critical nature of phase correction for the accurate determination of line positions and shapes. The asymmetric line shapes in Fig. 7.4c result from a 50:50 mixture of the real instrumental line shape (the sinc function) and the imaginary (the cosinc function). Phase correction minimizes the imaginary parts of the complex spectrum and consequently eliminates a majority of the line shape distortion as illustrated in Fig. 7.4d.

These figures show that phase errors can seriously degrade the spectrum, even in this case of an equal-sided interferogram. For unequal- or one-sided interferograms the effects are far more severe. It will be shown in Section 7.4 that the frequency shift of an emission line per radian of phase error is one-half its full width at half maximum, for a one-sided interferogram. For a signal-to-noise ratio of 1000, the uncertainty in line position owing to noise is about 1/1000 of the line width. To reduce the phase shift error to the same level requires a phase correction accurate to 2 milliradians.

7.4 Asymmetric Truncation, Amertization, and Phase Errors

Truncation is the multiplication of the interferogram by a rectangular function, which cuts it off at well-defined limits. Two extremes are *symmetrical* truncation (equal-sided interferogram), which gives the best and most accurate spectrum when transformed, and *completely asymmetrical* truncation (one-sided interferogram). A one-sided interferogram offers maximum resolution for a given observation time, but in return it sets more severe restrictions on the accuracy with which the phase must be determined. It may be helpful to review a few facts from Section 6.4 on apodization.

When an interferogram is truncated while its amplitude is appreciable, apodization is frequently employed. It is the multiplication of an interferogram by a function that shapes or smoothes the envelope of the interferogram, especially at the beginning and end. Its purpose is to eliminate sharp changes in amplitude that produce unusual spectral line shapes or satellites in the transformed spectrum. For example, when the maximum path difference is too small for spectrum lines to be fully resolved, the visibility of the interferogram fringes is still appreciable at the ends, where they stop abruptly. Apodization can remove the resulting false lines — satellites at the feet of strong lines, called *ringing*. In cases where the resolution is more than adequate, the interferogram is self-apodizing, as illustrated earlier, because the amplitude of the fringes decreases with increasing path difference until the contrast becomes negligibly small, and it makes no sense to keep taking data. The wasted resolution adds only noise.

Asymmetric truncation together with apodization are frequently encountered because the interferogram is not symmetrical to begin with. This asymmetry compounds the difficulties of obtaining a true representation of the spectrum. The larger the asymmetry, the larger the imaginary part of the transform and the consequent larger accuracy required in the phase correction. The functions used are often combined into one and called the *apodizing function*. The most accurate spectral profiles are obtained when we maximize the real part and minimize the imaginary part.

When we have only a one-sided interferogram, we can still minimize the effects of asymmetries. We separate the interferogram into even and odd parts by constructing suitable functions that are used in the same manner as apodizing functions. We apply these functions to the interferogram to construct a new interferogram that will reproduce as closely as possible the spectrum as if the interferogram were symmetrical.

To illustrate the problems encountered with asymmetric envelopes, let's look first at an interferogram, then at a particular apodizing function appropriate for this interferogram, and finally at the function's transform, which is called the *instrumental function* (the frequency response to an infinitely long sinusoidal wave input to the spectrometer). The process is often called *amertization*, after Larry Mertz.

Consider the unequal-sided interferogram shown in Fig. 7.5a. The sharp cutoff at the left will introduce artifacts into the spectrum. Also, low frequencies out to the path difference at the negative cutoff point will be twice as large in amplitude as those at larger path differences (two datum points at each interval, right and left, rather than only one at the right). Further, the nonzero average amplitude will introduce serious errors in the transform.

The solution is first to construct the function in Fig. 7.5b, which gives double weight to the unbalanced part of the scan, as if the unbalanced part were folded over to make an equal-sided interferogram, and smoothes the discontinuities. It can be decomposed into an even part that extends equally about the origin as if the interferogram were equal sided, and into an odd part that is zero near the center, where the fringes are strongest.

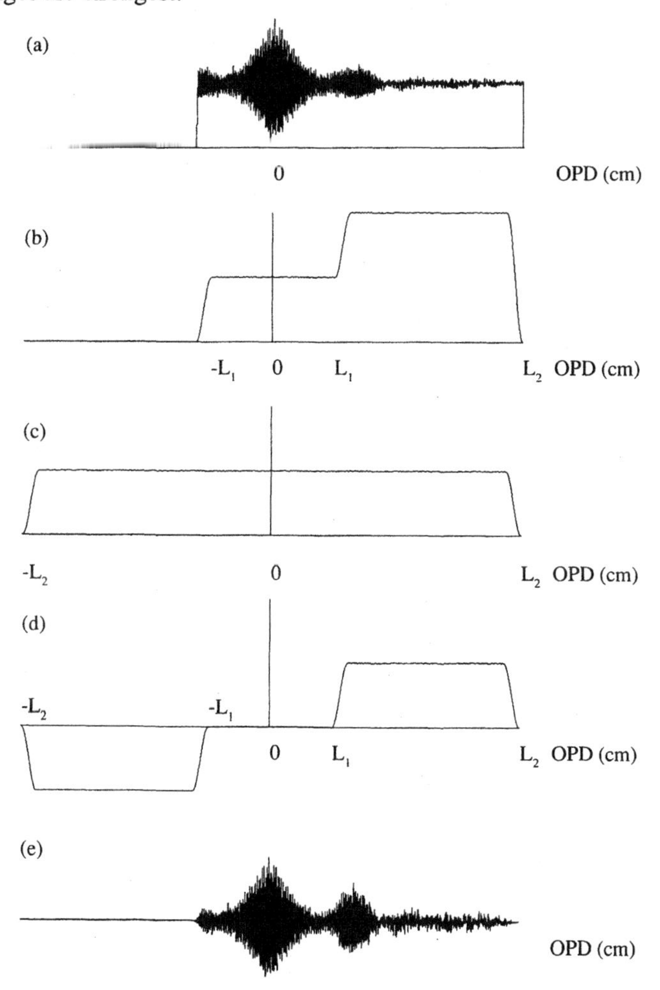

Fig. 7.5 (a) Unequal-sided interferogram (1:3 asymmetry). (b) Amertization function, which can be decomposed into even and odd parts. (c) Even part of function. (d) Odd part of function. (e) Apodized interferogram (zero corrected).

The two sections of the envelope are joined and terminated by short, curved cosinusoidal sections that ensure that both the function and its derivatives are free of discontinuities, as shown in the figure. The final step is to multiply the interferogram by the amertization function and to subtract off the average value. See Fig. 7.5e.

The even part of the interferogram envelope (the even part of the amertizing function) transforms into $t_s(\sigma)$, a real symmetric sinc function, while the odd part transforms into $t_a(\sigma)$, an imaginary antisymmetric cosinc function.

$$t_s(\sigma) = L\text{sinc}\,(2\pi\sigma L) = L\frac{\sin\,(2\pi\sigma L)}{2\pi\sigma L} \tag{7.6}$$

$$t_a(\sigma) = L\,\text{cosinc}\,(2\pi\sigma L) - l\,\text{cosinc}\,(2\pi\sigma l), \tag{7.7}$$

where by analogy with the sinc function,

$$\text{cosinc}\,(2\pi\sigma L) = \frac{\cos\,(2\pi\sigma L)}{2\pi\sigma L}. \tag{7.8}$$

To illustrate the foregoing, Fig. 7.6a shows the real part of the transform for a well-resolved gaussian line. Figures 7.6b and c show the imaginary part of the transform for different values of the unequal-sidedness.

These asymmetric transforms introduce asymmetries and, consequently, position shifts for all spectral lines. These position shifts are more severe for sinc profiles than for gaussian profiles. They cannot be apodized away! Any method for recovering the true spectrum must use both real and imaginary parts of the instrumental function.

To give a more quantitative estimate, for a one-sided interferogram the frequency shift $\delta\sigma$ per radian of phase error of an emission line is approximately half the full width at half maximum:

$$\delta\sigma = 0.5W\;\delta\;(\text{rad}), \tag{7.9}$$

where W is the FWHM in cm^{-1}. The expression for the wavenumber error in determining the center of a well-resolved symmetric emission line in the presence of noise is

$$\delta\sigma = \frac{0.5W}{(S/N)}, \tag{7.10}$$

where *S/N* is the signal-to-noise ratio. From these two expressions we conclude that if the wavenumber precision is to be determined by S/N rather than the phase error, we require

$$\delta \text{ (rad)} < \frac{1}{(S/N)}. \tag{7.11}$$

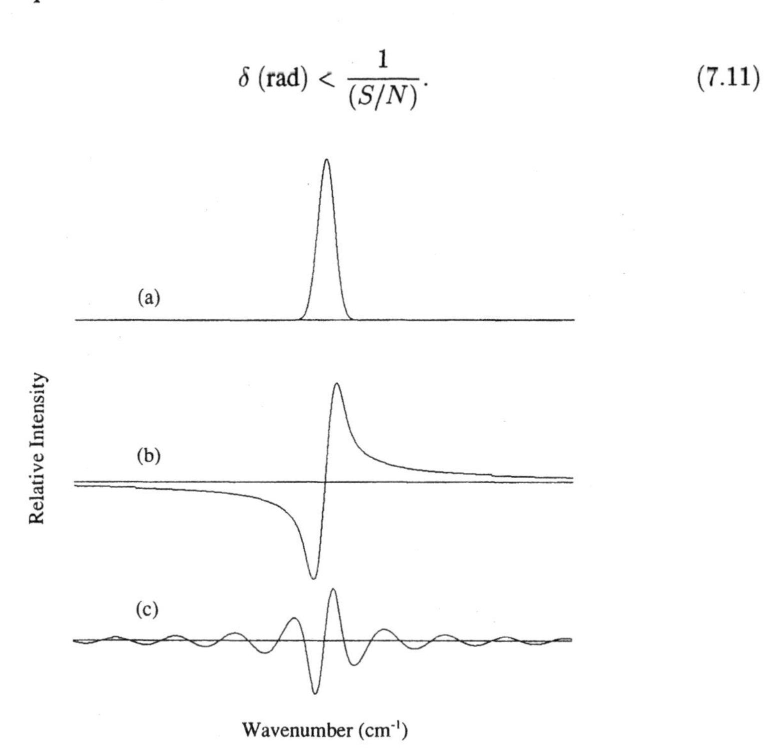

Fig. 7.6 The real and imaginary parts of the transform for a well-resolved gaussian line: (a) The real part of the line; (b) The imaginary part of the line for $L_1 = 0$; (c) $L_1/L_2 = 1/4$.

It should not be forgotten that truncation and apodization, whether symmetrical or asymmetrical, have direct effects on line widths and shapes. As shown in detail in Chapter 5, the transform of the truncating function is a filter or smoothing function that is convoluted with the true spectral line shapes. Too narrow a function makes the lines ring, unless the ends of the function are adequately smoothed by an apodizing function, but when they are so smoothed, the lines are excessively broadened and their true shapes obliterated. Too wide a truncating function reproduces line shapes accurately but introduces excess noise.

7.5 More On One-Sided vs. Symmetric Interferograms

Let's consider briefly the one-sided interferogram, since it is widely used. A common practice is to start the interferogram near zero path difference and observe it mainly on one side of the central fringe, with only a small fraction on the other side. Many instruments are incapable of recording a fully symmetric two-sided interferogram. The justification of this practice is simple: Both halves contain the same information, and if only a limited total path difference is mechanically available, the spectrometer yields twice the resolution when used in the one-sided mode. Alternately, only half as many numbers need to be recorded for a given resolution. Why, then, should symmetric interferograms be considered at all?

There are two major advantages to symmetric sampling of the interferogram:

1. All path differences are sampled symmetrically about a common mean time, so to first order all frequency components refer to the same mean epoch. This results in considerably more accurate line profiles in situations where the source intensity varies monotonically during an observation. Examples are astronomical sources near the times of rising or setting, when the air mass is rapidly changing, and laboratory sources that decay as some essential component is consumed or otherwise lost.

2. The sensitivity of line positions to the accuracy of the phase correction is drastically reduced. For one-sided interferograms, even a very small phase error is enough to rotate a portion of the imaginary part of the instrumental function into the real plane, producing an asymmetric instrumental function in the final spectrum. Since the phase error is often a function of wavenumber, this produces a varying instrumental function and hence a variable wavenumber scale. With two-sided interferograms, the only imaginary part comes from any slight asymmetries in the interferogram itself. Because these are normally very much smaller, the sensitivity to phase error is proportionally reduced. These facts are illustrated in Fig. 7.7. Needless to say, in an asymmetric interferogram, always take as many points as possible on the short side of the interferogram in order to minimize the effects of phase errors.

7.6 Determining Phase Shifts

In any event, whatever the source of phase shifts, the net result is that the position of zero path difference for each cosine wave in the interferogram depends upon its frequency. No single origin simultaneously exists for all frequencies. Hence we must first *determine* the phase shift function and then *correct* the interferogram itself or the resulting spectrum to compensate for its effect. Let's see how we can determine phase shifts in order that we may correct for them. When corrected, the interferogram is symmetric so far as the desired spectrum is concerned, although asymmetric for the noise.

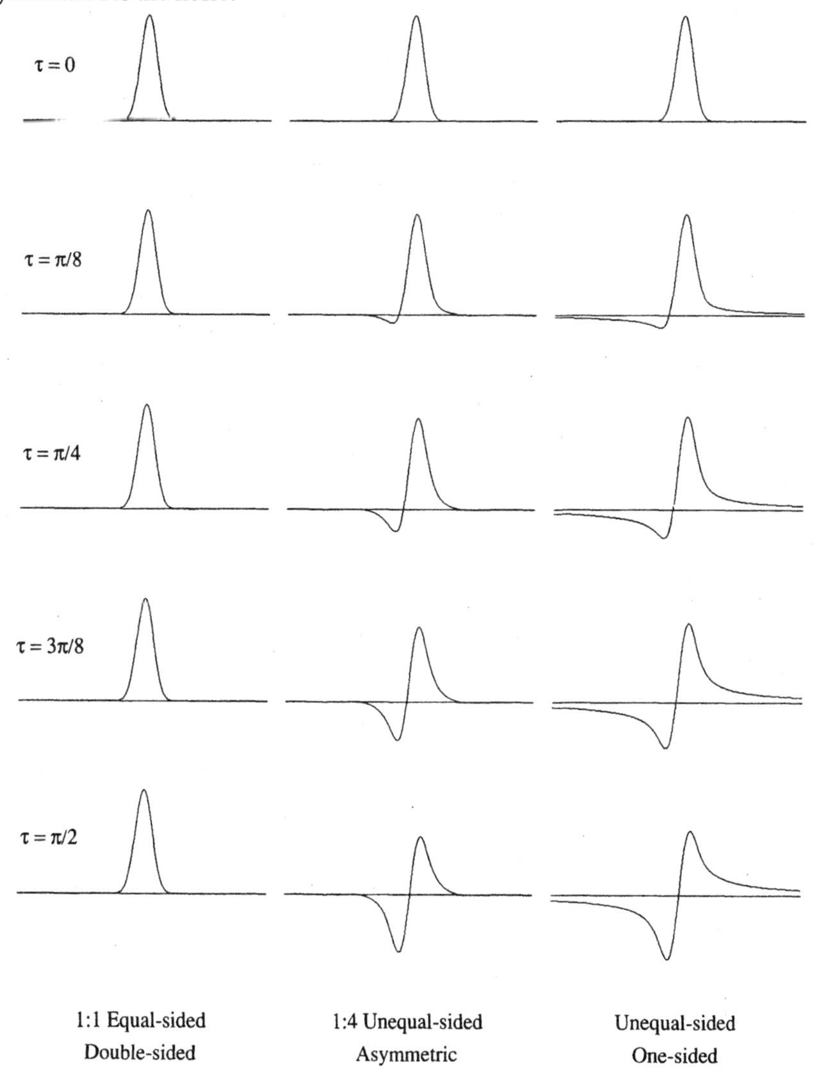

Fig. 7.7. Deformation of the real apparatus line shape function as a function of phase errors between 0 and $\pi/2$.

Consider an observed interferogram $I(x)$. For our present purposes, let us assume that it exists for a usefully large distance on both sides of the origin. We also assume that the origin has been chosen as close as possible to the point of zero phase, or at least to a point of "stationary phase." At this point, the imaginary part of the transform is at a minimum and the slope is near zero with respect to shifting the origin in either direction. Another way of saying it is: "If your original choice of center gives a steep phase plot, you move the center until it flattens out" [A. Thorne]. A poor choice here results in residual errors in line intensities and shapes even after a phase correction is made. The stationary phase point is often *not* at the point of maximum amplitude. Figures 7.1 through 7.3 show the foregoing, although the point shifts are in one direction only.

7.6.1 Low-Resolution Phase Determination

To determine the actual variation of the phase with frequency, the usual procedure is as follows. We assume that the phase varies only slowly with frequency so that a very low-resolution interferogram is sufficient to determine it. We truncate a large two-sided interferogram by multiplying the central portion by some symmetrical function $T(x)$ that falls smoothly to zero in a short distance (e.g., a cosine bell). Typically, 128 to 1024 points about the central fringe are used. The transform of this truncated interferogram yields a low-resolution phase nearly linear with wavenumber, a wide and smooth function that is assumed to be a good approximation to the true phase. It is then interpolated and used to correct the full-resolution transform.

The process of phase correction turns the complex spectrum $B(\sigma)$ into a real function by multiplying it by the rotation function $e^{i\phi(\sigma)}$. If it should happen that the phase is *not* a smoothly varying function of frequency, as occurs in the infrared, where thermal radiation from elements in the spectrometer mingle with source radiation, then we should take a medium-resolution interferogram, treat it as just described, take the transform, and then smooth it as necessary.

At this point we might look at some examples of when the phase varies smoothly and when it does not. An absorption spectrum is illustrated in Fig. 7.8 with the associated phase spectrum $\phi(\sigma)$. We can see that where the signal is large relative to the noise, the phase is well determined, but where the signal is small, as at the ends of the bandpass region or at the center of a very strong absorption line, or between strong emission lines the phase is not smooth.

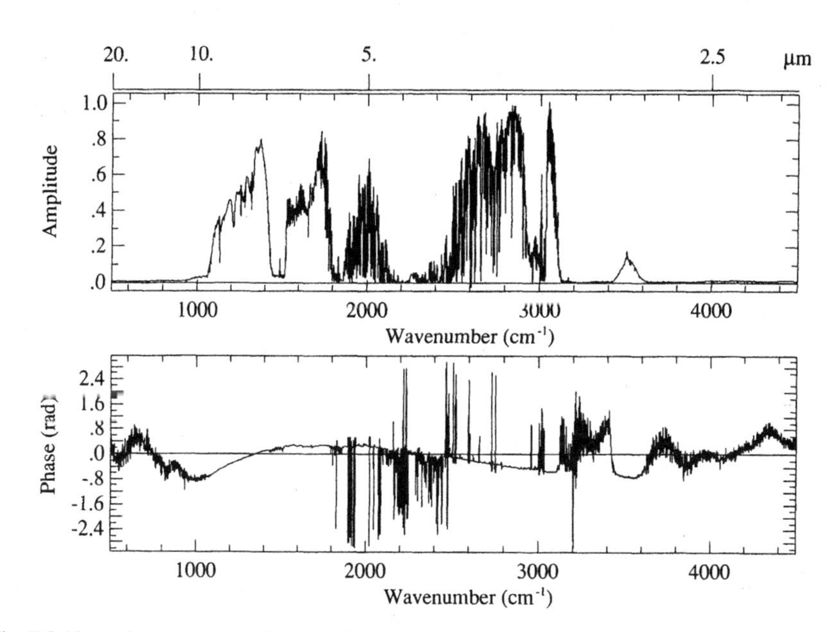

Fig. 7.8 Absorption spectrum and measured phase spectrum.

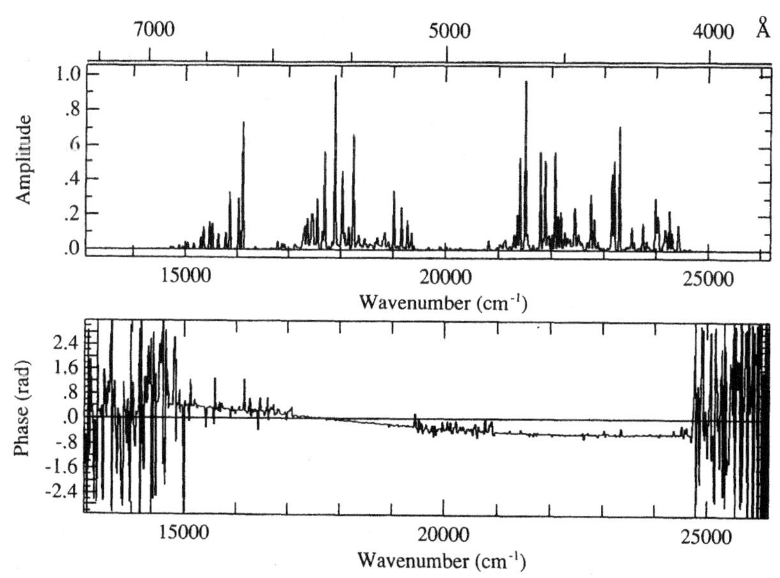

Fig. 7.9 Emission spectrum and measured phase spectrum.

In the case of an emission spectrum with only a few strong lines, the phase is well determined only at a few well-separated regions where the lines are, as shown in Fig. 7.9. There is no smoothly varying continuous function for $\phi(\sigma)$. The points must be fitted to a polynomial to get a smooth function. If the fitting is not done properly, the derived spectrum may exhibit bizarre behavior, with many unphysical oscillations, and consequently may be unrecognizable.

7.6.2 High-Resolution Phase Determination

There is a special problem posed by emission spectra — the propagation of the phase of strong lines into other regions by a too-narrow apodizing function that produces long wings on underresolved narrow spectral lines.

The obvious solution is to use a wide truncating function that includes many points in the interferogram around the center burst. After transformation, the resulting phase spectrum can be fit with a polynomial, using only those points in the spectrum that have sufficient amplitude to yield a reliable phase estimate. A procedure of this kind seems to be essential for dealing with a phase function that is well defined only in narrow isolated ranges, unless such ranges are numerous and well distributed. The main improvement will always be a better set of functions for the least-square fit. Polynomials are notoriously ill-behaved near the edges of the range over which the data are fitted, and extrapolation is worthless. Sometimes it is useful to take a physically more meaningful approach and *model* the sources of phase shift and to use as few adjustable parameters as possible.

7.7 Recommendations

From this intermediate-level treatment of the phase correction problem we can formulate some recommendations that we have used to advantage in our work. The message should be loud and clear: *Always* correct the phase, no matter how good the spectrum appears to be without a correction.

1. Minimize ϕ' — do not simply take the point of maximum amplitude as the center of the interferogram. Seek at least a point of stationary phase. In examining the interferogram, just as in fitting the spectrum, avoid linear interpolation between observed points [see the discussion in Chapter 6 beginning with Eq. (6.4)].

2. *Smooth* the phase by fitting it to a polynomial function or perhaps a function derived from a model.

3. Always take as many points as possible before the center burst — symmetric interferograms are best from the point of view of phase errors, especially with emission spectra.

4. Test experimentally the validity of results that may be affected by phase errors, by varying the scan parameters in some significant way and recalculating the spectrum. Systematic errors often can be uncovered because they are reproducible.

8

EFFECTS OF NOISE IN ITS VARIOUS FORMS

We have been assuming that the interferogram that we actually observe has no random variations in amplitude or phase that would appear in the spectrum as noise — sharp spikes, false spectral lines, or other features with no predictable basis in the incoming radiation.

In practice there are always various kinds of noise in the spectrum, generated either in the source or by electrical and mechanical variations in the environment or in the FTS itself. Most often the noise is not discernable in the interferogram, but appears clearly and looks like sharp spikes in the spectrum with random positions, widths, and intensities. It is "grass" around the zero line. It is true, however, that noise can also be periodic, as in the case of electrical pickup in the detector of 60-Hz electric fields from nearby power wires. And even if the equipment were perfect and introduced no noise, there would still be quantum or photon noise inherent in the radiation itself. To see some of the practical consequences of noise in FTS and how to avoid them, you may wish to turn now to the summary at the end of the chapter (Section 8.6), before trying to slog through the details.

It is helpful to note *how* various sources of noise and artifacts enter into the data. Source variations in frequency and intensity during the observation *multiply* the envelope of the interferogram in an undesirable way. Sampling position errors *add a term* to $I(x)$, depending on the *derivative* of the interferogram. Nonlinearities anywhere in the detection system *add on higher powers* of $I(x)$. Photon statistics and thermal processes in the detector and electronics *add* noise to the ideal interferogram. Fortunately, the errors do not usually interact, and we can consider each of them separately.

8.1. Signal and Noise in the Two Domains

For both historical and practical reasons, it is convenient to treat the noise in terms of where in the system it arises. Our work at the instrument is done in the interferogram domain. The real physical noise is generated there and the signal is measured there. Consequently, it is important to have a feeling for how procedures, processes, and events at the instrument translate into the final spectrum.

A wonderfully simplified analysis results when we work with a two-output interferometer, as discussed in Chapter 3 and diagrammed in Fig. 3.1. The interferogram is the difference of the two outputs, and when the two signals A and B have *uncorrelated* r.m.s. errors ϵ_A and ϵ_B such as might be generated by a jitter of one or both of the moving mirrors, then

$$I(x) = I_A(x) - I_B(x) \pm \sqrt{\epsilon_A^2 + \epsilon_B^2}. \tag{8.1}$$

The sum of the intensities of the two output beams is always the same as the total intensity of the incident beam, assuming no losses in the optics:

$$I_0 = I_A(x) + I_B(x) \pm \sqrt{\epsilon_A^2 + \epsilon_B^2}, \tag{8.2}$$

and because $I_B(0) = 0$ at the central fringe,

$$I_0 = I_A(0) = I(0). \tag{8.3}$$

These arguments say that the noise contribution to any sample in the interferogram, $\epsilon_x \equiv \sqrt{\epsilon_A^2 + \epsilon_B^2}$, is *independent of* x and has the same value as ϵ_A at the central fringe. A simple measure of the *signal-to-noise ratio* in the interferogram, one that is easily determined experimentally, is therefore

$$(S/N)_x = \frac{I(0)}{\epsilon_x}. \tag{8.4}$$

To go from a signal-to-noise measurement in the interferogram domain as just described, to a corresponding value for S/N in the spectral domain, we need to develop a mathematical relationship between the noise in the two domains.

First we make the assumption that, with few exceptions, the noise we are considering in this section is *white*, noise with equal power at all frequencies. Even if this assumption is not absolutely correct, as when we encounter a mixture of white noise and 1/*f* noise, it is usually adequate over the limited bandwidth that we commonly use.

To get the total r.m.s. noise power in the interferogram, we multiply ϵ_x^2 by the range $2L$. The transform of white noise is also white noise (refer to Fig. 4.16). Let the total r.m.s. noise amplitude in the spectral domain, including both real and imaginary parts, be $\epsilon_{\sigma,\text{tot}}$ and then the total noise power in the spectral domain is $\epsilon_{\sigma,\text{tot}}^2 2\sigma$. By the Rayleigh theorem (Eq. 4.45) the total powers in the two domains are equal, and

$$\epsilon_{\sigma,\text{tot}} = \epsilon_x \sqrt{2L/2\sigma_{\max}}. \tag{8.5}$$

But half of this noise power is in the imaginary part (the transform of the *odd* part of the noise), and we are interested in only the real part. If ϵ_σ is the real part, then

$$\epsilon_\sigma = \epsilon_{\sigma,\text{tot}}/\sqrt{2}. \tag{8.6}$$

Note that the noise in the spectrum is *uniformly distributed and does not depend on the local signal strength.*

Next we seek an equally simple relation between the *signals* in the two domains. We use the area theorem (Eq. 4.51),

$$f(0) = \delta\sigma \sum_n F(\sigma_n) \tag{8.7}$$

which in this case becomes

$$I(0) = \delta\sigma \sum_n B_e(\sigma_n) = \delta\sigma \cdot N \cdot \overline{B_e} \tag{8.8}$$

where $\overline{B_e}$ is the mean signal in the spectral domain and N is still the number of points. By combining these results and applying the relation

$$N = \frac{2L}{\delta x} = \frac{2\sigma_{\max}}{\delta\sigma} = 2L \cdot 2\sigma_{\max} \tag{8.9}$$

we obtain

$$\begin{aligned}(S/N)_x &= \frac{I(0)}{\epsilon_x} = \frac{\delta\sigma \cdot N \cdot \overline{B_e}}{\sqrt{2\sigma_{\max}/2L} \cdot \epsilon_{\sigma,\text{tot}}} \\ &= \frac{N \cdot \overline{B_e}}{\sqrt{2\sigma_{\max} \cdot 2L} \cdot \epsilon_{\sigma,\text{tot}}} = \sqrt{N/2} \cdot \frac{\overline{B_e}}{\epsilon_\sigma}.\end{aligned} \tag{8.10}$$

Finally, the *local* S/N ratio in the spectrum is

$$(S/N)_\sigma = \frac{B(\sigma)}{\epsilon_\sigma} = \sqrt{2/N}\frac{\cdot B(\sigma)}{\overline{B_e}} \cdot [S/N]_x. \tag{8.11}$$

So the relation between S/N in the spectral domain and in the interferogram domain is determined entirely by only two simple factors.

1. The number of points in the interferogram (or equivalently the maximum resolving power $R = N/2$). Higher resolution demands a greater S/N in the interferogram for a given S/N in the spectrum, although only as $\sqrt{N}$.
2. The ratio of *local* signal strength to the *mean* signal strength in the spectral domain. Therefore absorption spectra, and broadband spectra where there are both weak and strong signals, place much greater demands on the interferogram quality than do narrowband spectra or emission spectra. Another way of saying this is that the noise in the spectrum is proportional to the mean signal level only, in both absorption and emission.

8.2 Noise Classifications

Let us look at the noise in more detail. We will broadly classify noise sources according to their exponents in the equation $\epsilon_x = aI^k$:

$k = 0$: noise independent of source intensity (detector noise, digitizing noise)

$k = 0.5$: noise proportional to the square root of intensity (photon noise)

$k = 1$: noise proportional to intensity (source noise)

8.2.1 Detector Noise (k = 0): The Multiplex Advantage

Though less often encountered today, the case of detector noise-limited data was the one for which the FTS was invented and that motivated all of the early development of the technique. Consequently, it is well treated in the literature.

In short, if we can put the light from m spectral elements on the detector simultaneously, as we do with an FTS, *without increasing the noise,* then clearly the information flow is increased by that same factor as compared to a sequential single-element scanner such as a scanning grating. This is the so-called *multiplex advantage*.

In the very early days of the technique, when only relatively insensitive detectors were available in the infrared, this gain was sometimes as large as 1000. With modern detectors the situation has changed drastically, and we are now nearly always photon-noise limited. It may still be true, however, that an FTS observing, say, 10^5 spectral elements is photon noise limited with a given detector, while a sequential scanner looking at a single spectral element would be detector noise limited. In this case we should perhaps consider the FTS to have only a *partial multiplex advantage*.

8.2.2 Digitizing Noise ($k = 0$): Dynamic Range

The signal from the detector is an analog signal that must be converted into a digital one in order to process it. You may not wish to read all about it in the text that immediately follows unless you are designing an FTS, but the results are important for thinking about how to take data.

One conclusion is that for very simple emission spectra, digitizing noise is negligible, but the dynamic range of observable intensities varies inversely as the number of very strong lines. Another is that when the sampling rate is large enough that photon noise exceeds the digitizing noise, the coadding of n scans reduces the final noise by $\sqrt{n}$. For weak sources this is a perfectly feasible approach, but unfortunately is impractical for the Sun and other bright sources. However, if the initial low detector gain for bright sources is increased as the fringe amplitude decreases, the noise is reduced in direct proportion to the increase in gain.

For simplicity in analyzing the problem, let us return to monochromatic light and for clarity begin with discrete notation, although the notation is somewhat ambiguous because the symbol δ is used for both the Dirac delta function and an increment in σ. Let $B(\sigma) = \delta(\sigma - \sigma_o)$ so that

$$B_e(\sigma) = (1/2)[\delta(\sigma + \sigma_o) + \delta(\sigma - \sigma_o)] \tag{8.12}$$

and therefore

$$\begin{aligned} I(x) = \widetilde{B}_e &= \delta\sigma \sum_n B_e(\sigma_n) e^{i2\pi\sigma_n x} \\ &= \delta\sigma(1/2)[e^{i2\pi\sigma_o x} + e^{-i2\pi\sigma_o x}] \\ &= \delta\sigma \cdot \cos(2\pi\sigma_o x). \end{aligned}$$

We have a single sharp line of strength 1 in the spectral domain, which results from fringes of amplitude $\delta\sigma$ in the interferogram. Now we must specify the characteristics of our analog-to-digital converter (ADC), a crucial element in the system. Let

$$\begin{aligned} D_R &= \text{dynamic range of the ADC} = \frac{\text{full scale}}{\text{least bit}} \\ &= 2^n \text{ for a converter with } n \text{ bits, not including sign.} \end{aligned}$$

For the moment assume a perfect converter so that the only error comes from the fact that as the input monotonically increases, the digital output rises in discrete steps. The error as a function of input is therefore a triangular function with peak amplitude of one-half of the least significant bit (1/2 lsb). A calculation shows that the r.m.s. error in this case is $1/(2\sqrt{3})$ lsb, but we will take 1/3 bit for simplicity. What we care about is the digitizing error or noise expressed in terms of the dynamic range. For example,

$$\frac{\text{r.m.s. error}}{\text{full scale}} \sim 1/(3D_R) \sim 10^{-5} \text{ for } n = 15 \text{ bits.} \tag{8.13}$$

Now we can find the effect on the spectrum. If we adjust detector and amplifier gains so that the interferogram peaks just reach full scale on the ADC, then the digitizing noise is

$$\epsilon_x = \frac{\text{full scale}}{3D_R} = \frac{\delta\sigma}{3D_R}. \tag{8.14}$$

Random digitizing noise is essentially white (gaussian-distributed) noise, and hence its transform is also white noise. From the power theorem we know that the spectral white noise then has an r.m.s. value of

$$\epsilon_\sigma = \epsilon_x \cdot \sqrt{2L/2\sigma_{\max}} = \frac{\delta\sigma}{3D_R} \cdot \sqrt{2L/2\sigma_{\max}} = \frac{1}{3D_R\sqrt{N}}. \tag{8.15}$$

And since the line has an amplitude of 1/2 (we are still working with B_e, not B), the signal-to-noise S/N ratio there is $3D\sqrt{N}/2$. Finally, going from B_e to B improves S/N by $\sqrt{2}$, since half of the noise power is in the imaginary part. The final improvement is therefore a factor of $\sqrt{N/2}$, which is often of the order of 10^3. Now, this should not surprise us, since we made N measurements of the interferogram to determine an amplitude of 1 in the spectrum. Do not assume from this calculation, however, that we can therefore easily achieve $S/N \sim 10^8$! The only reasonable conclusion is that, for very simple spectra, digitizing noise is negligible. In fact, the dynamic range degrades roughly inversely as the number of strong lines in the spectrum.

Next, let us look at the other extreme: the common and astronomically interesting case of continuous absorption spectra. Let $B(\sigma) = 1$ for all σ, so $B_e(\sigma) = 1/2$ everywhere, and

$$\begin{aligned} I(x) &= \delta\sigma \sum_n B_e(\sigma_n) e^{i2\pi\sigma_n x} \\ &= \frac{\delta\sigma}{2} \sum_n e^{i2\pi\sigma_n x} = \frac{\delta\sigma}{2} N\,\delta(x) \end{aligned} \tag{8.16}$$

with $I(0) = N\,\delta\sigma/2 = \sigma_{\max}$, and with all other points in the interferogram are equal to zero. The proper units are implicitly contained in the choice of 1 for B. Again, if the gain is adjusted such that $I = \sigma_{\max}$ gives full scale, then $\epsilon_x = \sigma_{\max}/3D_R$ and

$$\epsilon_\sigma = \epsilon_x \sqrt{2L/2\sigma_{\max}} = \frac{\sigma_{\max}}{3D_R}\sqrt{2L/2\sigma_{\max}} = \frac{\sqrt{L\sigma_{\max}}}{3D_R} = \frac{\sqrt{N}}{6D_R}. \tag{8.17}$$

Since $B_e(\sigma) = \frac{1}{2}$, S/N is simply $3D_R/\sqrt{N}$, and now S/N for $B(\sigma)$ is $\sqrt{2} \cdot 3D_R/\sqrt{N}$. This time S/N is *reduced* by the factor $\sqrt{N/2}$ instead of being improved. Taking $N = 2 \cdot 10^6$ and $3D_R \sim 10^5$ (a 15-bit plus sign converter), we find an S/N of only 100 — clearly an unacceptable limitation for large-signal applications such as solar or laboratory absorption spectroscopy.

One solution to this problem is to sample at such a high rate that detector noise in the short samples exceeds the digitizing noise. Coadding n_s scans then reduces the final noise by $\sqrt{n_s}$. When sampling at 2500 Hz we sometimes have enough photon flux that S/N due to photons alone is nearly 10^6, yet we would have to sample 100 times faster to increase the detector noise to the level of the digitizing noise.

A different technique of quite general applicability is to use gain changes in the preamplifier, increasing the gain when the fringes become smaller. If the gain is increased by a factor of k_i for n_i points, then the digitizing error is *reduced* by k_i for those points, and the mean digitizing error is reduced to

$$\overline{\epsilon_x}^2 = \frac{1}{N} \sum n_i \left(\frac{\epsilon}{k_i}\right)^2. \tag{8.18}$$

In practice, this technique is very effective. In the most complex implementation, k_i is a power of 2, adjusted for each sample automatically. Simpler arrangements using powers of 4 or even 10, changed in blocks electrically or manually, can do quite well in many situations, especially if something about the range of values in the interferogram is known beforehand.

Finally, we should note that the problem is also reduced when we do not fill the whole bandwidth $\sigma_{\max}$ with information. If only a fraction α of the spectral elements is filled, the noise contribution to S/N from digitization is improved by $1/\alpha$.

8.2.3 Photon Noise ($k = 0.5$)

Fortunately, calculation of the noise due to statistical fluctuations in the rate of arrival of photons, "photon noise," is extremely simple: If n photons are collected on the average in one sampling time, the noise is $\pm\sqrt{n}$. Since our instrumental goal is always to reduce all other noise sources to the point where photon noise dominates, this result is relevant for any experimental observation.

Suppose we can count photons for a time τ seconds and that the average arrival rate is r photons/second. Then the total number collected is $n = r\tau$. Usually, however, the rate is too high to count individual photons, and we must resort to analog techniques followed by an ADC. With a primary unamplified photocurrent of i amperes, $r = i/e$, where e is the charge on the electron. Furthermore, if we are sampling the output at a rate f_s samples/second, the effective integration time is $\tau = 1/f_s$, so in one sample we get

$$n = r\tau = \frac{i}{e} \cdot \frac{1}{f_s}, \tag{8.19}$$

which gives

$$(S/N)_x = \frac{I(0)}{\epsilon_x} = \frac{n}{\sqrt{n}} = \sqrt{n} = \frac{\sqrt{i}}{\sqrt{ef_s}} \tag{8.20}$$

and the local S/N expected in the spectrum is (see Eq. 8.11)

$$(S/N)_\sigma = \frac{B(\sigma)}{\epsilon_\sigma} = \frac{1}{\sqrt{R}} \cdot \frac{B(\sigma)}{\overline{B_e}} \cdot (S/N)_x = \frac{\sqrt{i}}{\sqrt{ef_sR}} \cdot \frac{B(\sigma)}{\overline{B_e}}. \tag{8.21}$$

This is not a trivial result: It tells us that having chosen the resolving power R, and knowing something about the gross distribution of energy in the spectrum $(B/\overline{B_e})$, then:

> *A measurement of the total signal current is sufficient to predict S/N in the final spectrum in detail.*

This is an enormously comforting result, because as long as the resulting spectrum matches our expectations (our S/N predictions), we know that we have done as well as we could with the photons we are given to work with. The only way to improve the result is to find more photons somewhere.

8.2.3.1 Photon Noise ($k = 0.5$): Detector Efficiency

Spectral detectivities $D^*(\sigma)$ are commonly used to characterize the sensitivity of infrared detectors, but for most spectroscopic purposes they are quite misleading. Instead, *spectral detector quantum efficiencies* $\eta_q(\sigma)$ are needed in order to calculate the S/N ratio for the photon-noise-limited situation. Some of the arguments are complex and detailed. The results are summarized in Fig. 8.2, which you can use without reading further.

8.2.3.2 Spectral Detector Quantum Efficiency $\eta_q(\sigma)$

Let us start with a definition and description. The *spectral quantum efficiency* is defined as the ratio of the number of detected photons per second per wavenumber $\phi_D(\sigma)$ divided by the number of incident photons per second per wavenumber $\phi(\sigma)$:

$$\eta_q(\sigma) = \frac{\Phi_D(\sigma)}{\Phi(\sigma)}. \tag{8.22}$$

The signal output current for a *photovoltaic* detector is

$$I_s = q \int_0^\infty \Phi(\sigma)\eta_q(\sigma)\, d\sigma = q\Phi_D, \tag{8.23}$$

where Φ_D is the total number of detected photons per second, I_s is the detector signal current, and q is the charge of the electron. In the absence of any incoming radiation, the noise current is caused primarily by thermal noise in the feedback resistor of the detector preamplifier, known as Johnson noise. But for bright emission sources or for most absorption spectra where many spectral elements are filled, the r.m.s. noise current $I_{\text{r.m.s.}}$ is due mostly to statistical fluctuations of the photon flux and is therefore proportional to the square root of the detected photon flux:

$$I_{\text{r.m.s.}} = q\sqrt{2\,\Delta f\,\Phi_D} \tag{8.24}$$

where Δf is the frequency bandwidth (root 2 occurs because the sampling frequency is twice the bandwidth). We can then calculate the S/N ratio for the detector output if we know the detected photon flux and the frequency bandwidth:

$$(S/N) = \frac{I_s}{I_{\text{r.m.s.}}} = \sqrt{\frac{\Phi_D}{2\,\Delta f}} = \sqrt{\frac{\int_0^\infty \Phi(\sigma)\eta_q(\sigma)\, d\sigma}{2\,\Delta f}}. \tag{8.25}$$

For a *photoconductive* detector (rather than photovoltaic), the noise is larger by a factor of $\sqrt{2}$. When we use this equation to calculate the quantum efficiency from the S/N ratio for an ideal photoconductive detector, we obtain effective quantum efficiencies of $\eta(\sigma) = 1/2$.

8.2.3.3 Spectral Detectivity $D^*(\sigma)$

Data sheets for commercial detectors characterize the responses and sensitivities of detectors in terms of spectral detectivities:

$$D^*(\sigma) = \frac{\sqrt{A_D \Delta f}}{\text{NEP}(\sigma)}. \tag{8.26}$$

A_D is the detector area and NEP(σ) is the noise equivalent power. The D^* values are obtained for the background-noise-limited situation and are useful to give the S/N ratio using NEP directly when the background flux of room temperature or other thermal radiation is comparable to the photon flux of the source, an undesirable situation. This is the case for military applications, where there is a large blackbody background and only a very small signal. But in the laboratory it is almost like saying: "Hold a photomultiplier out in the room light to see what weak signal strengths you can detect."

In spectroscopy, the experimental setup suppresses the background flux by limiting the acceptance angle of the detector to the throughput solid angle of the instrument in order to optimize the sensitivity. We then need *spectral quantum efficiencies* rather than spectral detectivities to enable us to calculate the S/N ratio for the detector output, as in Eq. (8.25). These are *not* given on data sheets supplied by the manufacturers.

8.2.3.4 Conversion of $D^*(\sigma)$ to $\eta_q(\sigma)$

In order to calculate S/N for the background-limited case, we must know the background photon flux $\Phi_{B_D}(T_B)$:

$$\Phi_{B_D}(T_B) = \sin^2 \Theta_{1/2} A_D \int_0^\infty \eta_q(\sigma) M(T_B, \sigma)\, d\sigma, \tag{8.27}$$

where $\Theta_{1/2}$ is the half-angle field of view, A_D is the detector area, and $M(T_B, \sigma)$ is the photon flux exitance in photons/sec·cm^2·cm^{-1}. The last is given by:

$$M(T_B, \sigma) = \frac{2\pi c \sigma^2}{e^{[hc\sigma/kT_B]} - 1}, \tag{8.28}$$

where c is the speed of light, h is Planck's constant, and T_B is the temperature of the background. Then S/N for a given probe signal is:

$$S/N = \frac{\text{probe signal}}{\text{background noise}} = \frac{\Phi_S(\sigma)\eta_q(\sigma)}{\sqrt{2\,\Delta f\,\Phi_{B_D}}}, \tag{8.29}$$

where $\Phi_S(\sigma)$ is the photon flux of the probe source.

When we set $S/N = 1$ in this equation we get the noise equivalent photon flux. Assuming that all signal photons have the same energy, we obtain the NEP by multiplying the noise equivalent photon flux by the energy of the photons $hc\sigma$:

$$\text{NEP}(\sigma) = hc\sigma\Phi_S(\sigma)|_{S/N=1} = \frac{hc\sigma}{\eta_q(\sigma)}\sqrt{2\,\Delta f\,\Phi_{B_D}}. \tag{8.30}$$

Using the definition of $D*$ given in Eq. (8.26), this can be rearranged to give

$$\eta_q(\sigma) = hc\sqrt{\frac{2\Phi_{B_D}}{A_D}}D^*(\sigma) = k\sigma D^*(\sigma), \tag{8.31}$$

where k is a constant to be determined. Thus we see that the *shape* of $\eta_q(\sigma)$ is the same as that of $\sigma D^*(\sigma)$, while the constant of proportionality depends on the integral over the whole thermal region, which can be written

$$k^2 = \frac{2h^2c^2\Phi_{B_D}}{A_D} = 2h^2c^2\,\sin^2\,\Theta_{1/2}A_Dk\int_0^\infty \sigma D^*(\sigma)M(T_B,\sigma)\,d\sigma. \tag{8.32}$$

Canceling one power of k and substituting into Eq. (8.31), we get the desired formula

$$\eta_q(\sigma) = 2\sigma D^*(\sigma)h^2c^2\,\sin^2\,\Theta_{1/2}\int_0^\infty \sigma D^*(\sigma)M(T_B,\sigma)\,d\sigma. \tag{8.33}$$

To illustrate the use of this formula, we have calculated the spectral quantum efficiencies for the photovoltaic detectors indium antimonide (InSb, T = 77 K) and indium arsenide (InAs, T = 77 K) and the photoconductors mercury-cadmium-telluride (MCT, T = 77 K) and copper doped silicon (Si:Cu, T = 4.2 K) from D* curves (see Fig. 8.1) given in the familiar wall chart distributed by detector manufacturers. The resulting curves are given in Fig. 8.2.

Several facts can be noted. The InSb detector has the highest quantum efficiency of all the listed detectors and is nearly three times more sensitive than the InAs detector, even though the latter has a *much* higher D* value. The "wideband" MCT detector, which has been widely used for FT-IR and laser spectroscopy, has a very poor quantum efficiency compared to that of the Ge:Cu detector. Although narrowband MCT detectors have now become available, it is a historically relevant comparison.

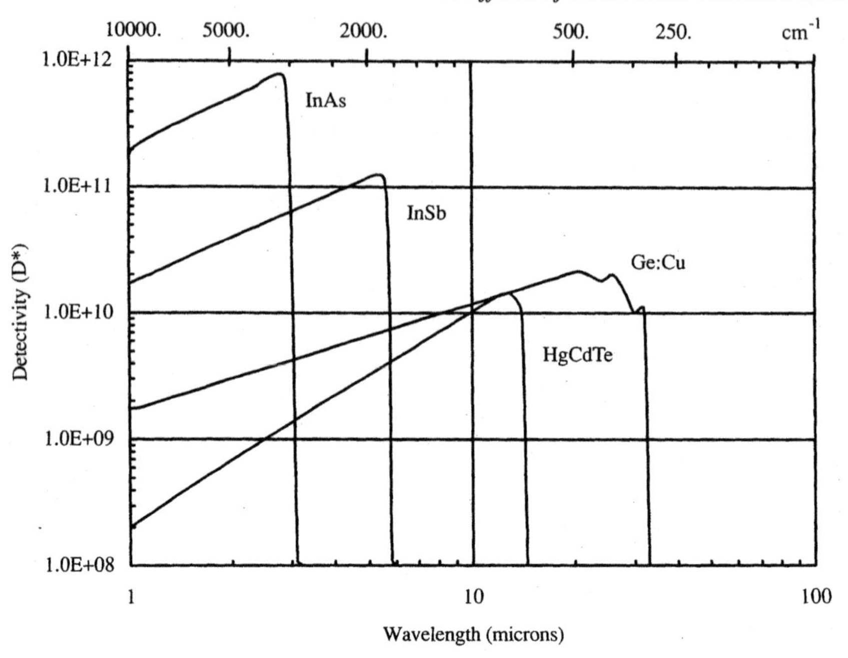

Fig. 8.1 Spectral Detectivities $D^*(\sigma)$.

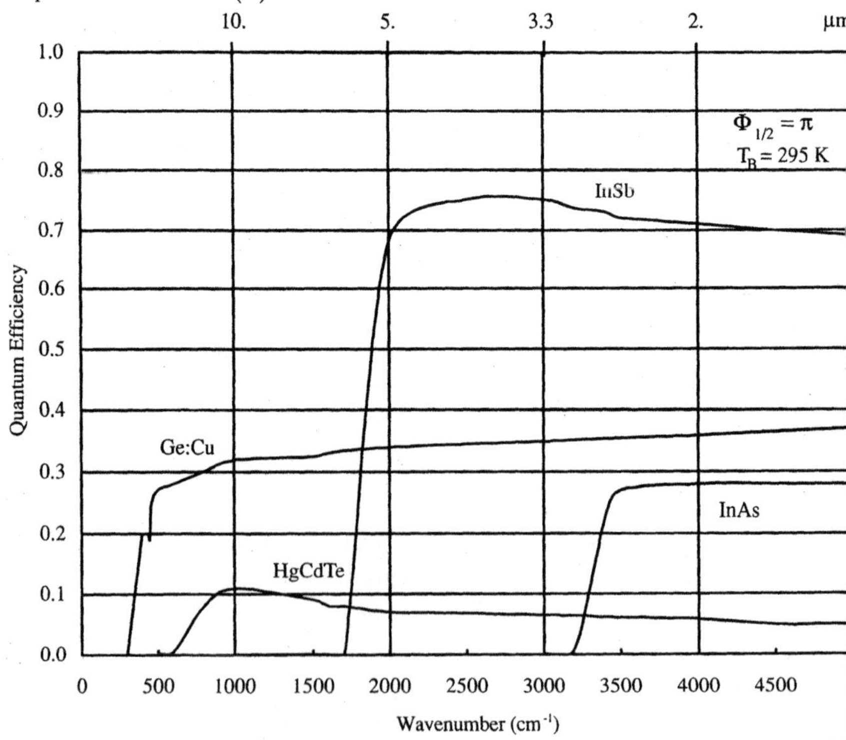

Fig. 8.2 Spectral detector quantum efficiencies $\eta_q(\sigma)$.

8.2.4 Source Noise ($k = 1$)

Source noise arises from variations in intensity with time which may be periodic as well as random in character. Since $k = 0$ leads to a multiplex advantage and $k = 0.5$ is neutral, we might expect $k = 1$ to yield a multiplex *disadvantage*, and this is indeed the case. When we observe m elements simultaneously, the information flow is m times greater than for a single-element scanner. But since the noise is also up by m, we must observe a factor of m^2 longer to get the same S/N. This penalty can be severe, so every possible means should be used to reduce its effect.

Source noise enters the result in two ways. *Additive* noise appears through the constant term in the interferogram (Eq. 3.4), and *multiplicative* noise appears because the whole instantaneous interferogram is proportional to the instantaneous source intensity.

Additive source noise is simply any fluctuation in the source that appears at the same frequency as the fringes due to the observed signal. It dominated the thinking of many of the early workers in the field, because at that time fringe frequencies were relatively low (hertz or low tens of hertz) for technical reasons, while most sources exhibit a strong component of noise that varies as $1/f$ in that frequency range; this noise is therefore referred to as $1/f$ noise.

One solution to this problem is to use *internal modulation,* in which the path difference is rapidly switched back and forth over a distance of the order of half a fringe at a frequency higher than the $1/f$ noise, typically at a few hundred hertz. When this signal is synchronously demodulated and integrated, an output roughly proportional to the derivative of the interferogram is obtained.

Another solution adopted in most recent designs is the *rapid scan* technique, in which the path difference is simply varied fast enough to bring all fringe frequencies above the $1/f$ noise, usually frequencies in the 100- to 10,000-Hz range. No separate modulator is required because the FTS itself *is* a modulator, and hence no demodulation is needed. The additive $1/f$ noise is not suppressed, but merely appears in the spectrum at small wavenumbers, where there is no observable signal.

In most situations involving natural sources, one or the other of the preceding techniques (it is sometimes a matter of style which is chosen) is sufficient to deal with additive noise, because the $1/f$ component dominates. This is not always true of *laboratory* sources, which are sometimes subject to a variety of plasma oscillations in the audio range or above.

Next, *multiplicative* noise results because the source intensity also multiplies the cosine term in the expression for the fringe amplitude (Eq. 3.4). Each spectral element is convolved with an apparatus function that is the *Fourier transform of the intensity of that spectral element during the scan.* Now, since both outputs are proportional to the source intensity, and their sum sometimes *is* the source intensity, it is often suggested that this problem can be eliminated by taking the ratio of the difference and the sum as the signal; *i.e.*, if A and B are the two output signals, use

$$I(x) = \frac{A - B}{A + B}. \tag{8.34}$$

This looks good on paper, and in fact usually does *some* good in practice, but not as much as we would hope, for several reasons. First, the variation of the intensity is usually *wavenumber dependent,* so $A + B$ represents only some mean behavior and not the detailed spectral information that is really required. As examples, scintillation in the atmosphere is different for red and blue light, and carrier gas and metal lines behave differently (often oppositely!) in discharge lamps. Second, if there is any absorption in the beamsplitter, $A + B$ is *not* just the source intensity, because the fringes in the two outputs are no longer completely complementary. Finally, it is difficult to match the signal detection characteristics of A and B exactly. And if there is any fringe leakage in $A + B$ due to beamsplitter absorption, the frequency response of the $A + B$ signal *must* be very different from that of $A - B$, so the ratio is effective only for the lowest wavenumbers.

However, there is one approach that always helps, and that is the rapid scan technique. Suppose that we have one hour to make an observation, and imagine any plausible variation of the source intensity during that time. If we take one slow scan (as is usually the case for the step-and-integrate plus internal modulation approach), then the Fourier transform of the source variation is the instrument function. If the $1/f$ noise characteristic extends to frequencies as low as 1 millihertz, we may expect considerable power in the "grass" everywhere after transformation, due to this low-frequency noise. To put it another way, the r.m.s. variation of the envelope about its mean is likely to be a considerable fraction of the mean, and this translates into a poor signal-to-noise ratio in the instrument function.

Now, suppose instead that this same hour is used to scan, say, 16 times, with the 16 scans being coadded point for point so that the effective integration time is exactly the same. We have now made two major gains: First, because the lowest frequency present (i.e., one period/scan) is 16 times higher, this noise amplitude is down by a factor of 16 if the noise scales as $1/f$; second, if the noise is incoherent, we gain another $\sqrt{16}$ because the phases are random in the separate scans. But the important fact is that *these gains hold at each point in the spectrum, even if the time variation is chromatic.* This is a major advantage of rapid scanning.

An obvious conclusion to be drawn from our considerations of noise is that we should observe only that optical bandwidth that is absolutely necessary for the problem being studied.

8.3 Unbalance and Misalignment, Modulation Efficiency, and Signal-to-Noise Ratio

Let us recall equations for the two outputs in response to monochromatic light of unit amplitude

$$\begin{aligned} E_A &= \eta_o \eta_b \left[\frac{1 + \cos(2\pi\sigma x)}{2} \right] \\ E_B &= \eta_o - E_A = \eta_o \eta_b \left[\frac{1 - \cos(2\pi\sigma x)}{2} \right] + \eta_o (1 - \eta_b). \end{aligned} \tag{8.35}$$

The optical efficiency η_o is a maximum of 100% when the mirrors are perfectly reflecting, and the beamsplitter efficiency η_b is a maximum of 100% when there is no absorption and exactly half the light is reflected and half transmitted.

We first note that the balanced (A) output is completely modulated, with the quantity in the brackets in Eq. (8.35) varying between 0 and 1 independent of the beamsplitter efficiency η_b. This is what is meant by a modulation efficiency of 100%. The unbalanced (B) output differs in two ways. The sign of the modulation is reversed in the first term, and a constant second term has been added that is proportional to $(1 - \eta_b)$. When we take the peak-to-peak modulation divided by the overall peak signal (= 1), we see that the modulation efficiency for this output is only η_b.

Under what conditions is it better to use the signal from the second output as well as the first? Taking $E_A - E_B$ doubles the desired signal, but the photon noise is also increased due to the added constant in E_B. For the photon noise to be less than doubled, which would be the break-even point, the average flux on the detectors must be less than four times that of the balanced output alone. The mean level in E_B, which is equal to $(1 - \eta_b/2)$, must be less than three times that of E_A, $(\eta_b/2)$, which gives $\eta_b \geq 0.5$ for the limiting case. This condition defines the useful range of a given beamsplitter. The equivalent limitation on the reflectance R is that it should be between 0.146 and 0.854, since $\eta_b = 4RT = 4R(1 - R)$. However, if the dominant noise is source noise, it may be worth using the second output for noise cancellation even if the photon noise is increased by more than a factor of 2.

Misalignment of the optics or inadequate flatness and parallelism of the optical surfaces produce more serious losses, by preventing complete interference between the two beams. In these cases, the *mean* levels of both E_A and E_B remain unchanged, while the amplitude of the cosine term is reduced. The visibility of the fringes $(E_{\mathrm{max}} - E_{\mathrm{min}})/(E_{\mathrm{max}} + E_{\mathrm{min}})$, called the modulation efficiency e_{m}, is a direct measure of the signal, It decreases in both outputs while the noise remains constant. S/N is therefore directly proportional to e_{m}, and the time required for a given S/N varies as $1/e_{\mathrm{m}}^2$. Letting the modulation drop to 70% requires twice the observing time to get the same S/N.

8.4 Noise-Limited Resolution

With any type of spectrometer, it is always true that excessive resolution is expensive and should be avoided, but it is not always so obvious as to when the resolution has become excessive. The situation is clear and simple in the case of the FTS: so long as the interferogram is detectable above the noise, the accuracy of the spectrum is improved by higher resolution. In fact, we may want to scan *slightly* past this point, since the noise is often representative of a wider bandwidth than that of the signal, and a small increase in resolution can make a substantial decrease in the distortion of line shapes (refer to Fig. 5.6).

But it is a mistake to go too far beyond this point, since we lose in two ways: We are spending time uselessly that could have been used to observe the regions that contain useful signals, and we are producing points that contribute nothing but noise to the whole spectrum. As a result, S/N drops as $1/R$ once we have passed the optimum resolution.

8.5 Localized Noise (Ghosts and Artifacts)

Historically, high-resolution Fourier transform spectroscopy has dramatically improved the measurement of spectra of atoms and molecules because it offers accurate line positions and nearly linear intensity measurements over a wide range of values. It is free from the many spurious lines and noise present in diffraction grating spectra. But in turn it generates its own distinct set of false lines that interfere with or complicate a spectrum. They result from instrumental errors peculiar to the FTS. Errors that produce effects that are spread broadly throughout the spectrum are generally thought of as simply another form of noise, but when the effect is relatively localized the situation changes. As usual, the inverse relation between widths in the two domains holds here, and any process producing a recognizable feature must exist throughout most or all of the interferogram.

If the effect of instrumental errors is linked to the frequency of spectral lines, as, for example, by a constant frequency offset from every line, it is called a *ghost*. Ghosts are produced by *multiplicative* processes — the interferogram is *modulated* in some way by the noise source. In general, modulation of the intensity produces symmetric ghosts, while modulation of the path difference produces antisymmetric ones. If an instrumental error is additive and hence independent of the spectrum, as, for example, with a spike corresponding to the power line frequency, it is classified as an *artifact*.

For convenience we shall divide ghosts or false lines into two broad classes: those that are linked to the laser fringes and are called *coherent*, and those that are *quasi-coherent*. The first class includes ghosts due to leakage into the detector of laser fringes from the control interferometer and those due to nonlinearities in the detection system. These ghosts are faithful replicas of an original line except for a shift in wavenumber and a change in scale and a possible sign change. They correspond to Rowland ghosts in diffraction grating spectra, which arise from small sinusoidal variations in grating spacing about an average value.

The second class of ghosts is generally formed by power-line-related phenomena such as direct pickup, source modulation, and motor vibrations, or by mechanical resonances or plasma oscillations of high Q. Such ghosts range from nearly coherent to broad bursts of noise.

Nonlinearities and ghosts are present in all spectrometers, but the overriding complication in Fourier transform spectrometry is that most ghost lines fit the combination rules used for determining energy levels. The ghosts must therefore be found and eliminated. They are completely fictitious and a product of the instrument used to observe the spectrum. In our own experience we observed that spectra of different hollow cathode lamps, such as Ni, Fe, and Co, had similar emission lines of nonstandard profiles, which we originally attributed to the carrier gas. Instead they proved to be ghosts. With hindsight it is now easy to elucidate the theory of ghost generation and to check for the presence of ghosts empirically.

8.5.1 Periodic Position-Error Ghosts

Since our knowledge of the current value of the interferometer path difference is always subject to small but finite errors, we will in general observe not at exactly the desired distance x_m but instead at the nearby position $x_m + \epsilon_m$, and will recover a slightly different observed spectrum. We assume that ϵ is a small fraction a of the sampling interval, expand the intensity function about the desired position, and find that a periodic error produces a positive ghost at $\sigma = \sigma_0 + \delta\sigma$ and a negative ghost at $\sigma = \sigma_0 - \delta\sigma$ with $\delta\sigma = (2\sigma_{\max})k/N$, where k is the number of periods in the total interferogram and N is the number of points. The amplitude of each ghost is a fraction a of the line amplitude, the same fraction of the sampling interval as the error in the interferometer position. If the periodic error is 10^{-3} of a sampling interval (which might be less than a nanometer in physical path difference), we may expect ghost intensities of about 10^{-3} of the parent line. Since emission spectra often show dynamic ranges in excess of 10^5, such ghosts would appear as relatively strong lines.

8.5.2 Coherent Ghosts

Coherent ghosts can result from modulation that has a phase coherent with the laser light used to control the carriage position.

8.5.2.1 Laser Ghosts

Coherent ghosts are produced by a spurious signal in either the position servo or the sampling circuitry, resulting in a phase modulation of the interferogram with a period of one wavelength of laser light. They are found at frequencies corresponding to the sum or difference of the laser frequency and the spectral line frequency, or the aliases of such frequencies. Let

s_l = number of samples per laser fringe

σ_o = spectral line frequency in cm^{-1}

σ_l = laser frequency (15 798.0025 cm^{-1} for He-Ne)

$\sigma_{\max}$ = cutoff frequency of the observed spectrum = $0.5 s_l \sigma_l$.

The ghosts appear at such frequencies as $(\sigma_o + \sigma_1)$, $(\sigma_o - \sigma_1)$, and $(2\sigma_{\max} - \sigma_o - \sigma_1)$.

8.5.2.2 Nonlinear Harmonic Ghosts

Further ghosts are produced when the interferogram is modulated in some way by nonlinearities in the detectors or detection electronics or even mechanical linkages, and as a result false lines are introduced into the spectrum as parasites of strong spectral lines. They are most commonly observed in sparse atomic spectra containing a few very intense emission lines. Any instrumental system has nonlinear transfer characteristics in some frequency and amplitude regime, and the nonlinearities generate harmonics and mix frequencies, which then appear as ghosts.

Two major electronic nonlinearities are slope discontinuities in range-changing amplifiers used in the spectrometer, and photomultiplier saturation effects, which reduce the output intensity for large photocurrents. A range-changing amplifier might consist of as many as eight amplifiers that switch settings when the intensity changes by a factor of 2. Such a device is optimized to match the gain at each change. But to maintain slope continuity is more difficult, and the failure to do so produces nonlinearities on the order of parts per thousand. The transfer function has sharp corners, which cause high-order mixing, because high-frequency components are necessary to describe the corners. A classic example is the introduction of harmonics when a cosine oscillation is clipped by a detector. The resulting nearly square wave can be analyzed into the superposition of signals at one, three, five, etc., times the fundamental frequency. By contrast, saturation effects suppress the high frequencies and cause low-order harmonic mixing.

8.5.2.3 Nonlinear Intermodulation Ghosts

The nonlinear characteristics of the transfer function, the electrical analog of the optical instrument function, mean that instead of having an intensity distribution in the recorded interferogram that is a linear superposition of signals of the form

$$a \cos(A) + b \cos(B), \tag{8.36}$$

the intensity distribution contains additional cross-terms of the form

$$[a \cos(A) + b \cos(B)]^2. \tag{8.37}$$

These terms appear at well-defined frequencies in the transform, and the resulting peaks in the spectrum are called *ghosts*. Terms of the first kind, $\cos^2(A)$, give rise to harmonics since they can be reduced into $\cos(2A)$, while terms of the second kind, $\cos(A)\cos(B)$, mix different frequencies and produce sum and difference frequencies $\cos(A+B)$ and $\cos(A-B)$. In practice the specific coefficients of the nonlinear transfer function that determine the ghost intensities are not especially useful, but the *patterns* of the line positions and shapes are the key to testing for the presence of ghosts.

To demonstrate this problem in its simplest form, suppose we have a spectrum consisting of two monochromatic lines, and we look at the normalized output of a single detector so that

$$I(x) = 1 + a\cos(2\pi\sigma_a x) + b\cos(2\pi\sigma_b x). \tag{8.38}$$

And let there be a nonlinearity somewhere in the system that can be expanded in a Taylor's series so that we observe not $I(x)$, but

$$O(x) = I(x) + k_2 I^2(x) + k_3 I^3(x) + \ldots. \tag{8.39}$$

The first term is the desired result, and the rest produce unwanted ghosts. For the moment, let us assume that k_3 and all higher coefficients are negligible and discuss the effects of the quadratic term.

The quadratic term produces three groups of frequencies. The first lines are at the *original* line positions, but with modified intensities. This group is present or absent depending on whether or not the constant term in the ideal original transform is still present when the nonlinearity is encountered. In two-output systems it is balanced out, and in most other systems it is filtered out before the analog-to-digital converter.

The second group contains the second *harmonics* of our two lines, so we must look for ghosts at 2σ and its alias $2(\sigma_{\text{max}} - \sigma)$.

The third group contains the sum and difference frequencies of the two lines — the *intermodulation* products. Note that these same three groups of lines will result from *each pair* of lines in a more complex spectrum.

This example was chosen primarily for its mathematical simplicity. Such ghosts may well be important on single-output systems, especially at high light levels, where most detectors begin to show nonlinearities, but they are rarely seen in spectra from two-output systems such as the McMath FTS at the National Solar Observatory, Kitt Peak. Instead, an analysis of some of the latter spectra shows the following pattern: If σ_a and σ_b denote the frequencies of two very strong lines somewhere in the observed spectrum, then lines of moderate strength at wavenumbers σ_o exhibit ghosts at

$$\sigma_0 \pm (\sigma_a - \sigma_b). \tag{8.40}$$

These ghosts are symmetric pairs, usually positive.

The lowest-order nonlinear term that can produce ghosts that involve *three* frequencies is the cubic. Such a term may appear in the detector response at high light levels such as occur in solar or laboratory absorption measurements. It may also appear at any light level, if it is due to nonlinearities in the analog amplifiers, or more likely if it is due to scale or zero discontinuities in the multiple-range analog-to-digital converters commonly used to improve dynamic range.

8.5.2.4 Quasi-Coherent Ghosts

This type of ghost has properties similar to laser ghosts, except that the frequencies are generally known in hertz rather than in cm^{-1}. They may appear in an alias, and can be caused by intensity or phase modulation. When we take s_l samples per laser fringe, at a rate f_s per second, then

$$\frac{\Delta\upsilon}{\sigma_{\max}} = \frac{f}{0.5 f_s}; \qquad \Delta\sigma = \frac{f}{f_s} s_l \sigma_l, \tag{8.41}$$

where f is the frequency in hertz of the responsible phenomenon. Note that the constant of proportionality between $\Delta\sigma$ and f depends on both s_l and f_s, both of which may be operational variables. If one or both can be changed, then these ghosts will move to new positions and they can be identified.

8.5.3 Ringing Ghosts

Sharp, intense spectrum lines often display $\sin(x)/x$ ringing. It is symmetrical about the line center and occurs more prominently at the smaller-wavenumber end of the spectrum, where the Doppler width is less and the lines are sharper. Each ringing peak has an FWHM that is roughly equal to the resolution width of the interferometer, usually about 1/4 to 1/3 of a genuine line width. In selecting spectrum lines with a line-picking computer code, if the intensity cutoff is set high enough to miss these side peaks, then the fitting program will find the best line profile for the central peak, although the residuals will look bad. But lowering the limit of line picking to include the sidebands and then fitting them individually worsens the fit to the main line (as it should, since the fitting is meaningless). The ringing can always be removed by reducing the resolution through apodizing the interferogram, but this is a useful means only if the entire spectrum shows ringing. A better approach is to include the instrumental function in the fitting process, which permits iterative removal of the sinc oscillations and the precise measurement of lines with widths smaller than the resolution width of the instrument. Ringing is discussed in detail in Section 5.3, and methods for its systematic removal are given in Section 9.4.7.

8.6 Summary

1. Scan as rapidly as possible to avoid bad consequences of source noise. Practically, this will move the desired electrical signals (corresponding to the spectrum) as far away from the undesired electrical signals generated by the detector and the source (which generally decrease as $1/f$). Observe only that bandwidth that is necessary for the problem being studied.

2. A two-output interferometer simplifies the analysis of noise effects. The noise in the interferogram is independent of the path difference, and a useful measure of the signal-to-noise ratio in the interferogram is equal to the intensity at the central fringe divided by the r.m.s. noise.

3. The noise in the spectrum is uniformly distributed and depends only on the mean signal strength, for both emission and absorption, although not for a presentation in transmittance. For an interferogram of length L and a given S/N, the spectral S/N decreases with the square root of the number of sampling points N or, equivalently, with the square root of the resolution.

4. There is an FTS multiplex advantage in noise reduction, since all spectral elements are recorded simultaneously.

5. Digitizing noise is negligible for simple spectra, but the dynamic range of observable intensities for emission spectra varies inversely as the number of very strong lines. When the sampling rate is large enough that photon noise exceeds digitizing noise, coadding of n scans reduces the final noise by $\sqrt{n}$. In contrast, digitization noise can plague observations of absorption spectra, when small features have to been seen against a broadband spectrum with a large mean signal.

6. For large photon fluxes that require analog techniques for detection, a measurement of the total signal current is sufficient to predict S/N in the final

spectrum. Such a measurement is a excellent way to evaluate how close the observing condition are to the ideal case for photon noise limited observations.

7. Use the detector with the best quantum efficiency. That efficiency depends not just on the product of wavenumber and spectral detectivity as given in manufacturers' data sheets, but on their integral over the whole thermal region as well.

8. The use of both outputs is usually but not always the best choice. A re-examination of the issues is required for each change in source and operating conditions.

9. So long as the interferogram is detectable above the noise, the accuracy of the spectrum is improved by higher resolution.

10. Ghosts produced by a well-designed instrument are seldom troublesome and then mostly in spectra with the highest values of S/N and resolution. In most cases they tend to be insignificant or obvious, such as ringing and 60-Hz modulation.

nections. [illegible] instrument [illegible] excellent way to evaluate how ele[illegible] [illegible] the [illegible] motion moves from [illegible] system.

[illegible] with the new orchestra [illegible] [illegible] [illegible]

[illegible]

[illegible]

[illegible]

9

LINE POSITIONS, LINE PROFILES, AND LINE FITTING

9.1 Introduction

In spectroscopy, the observer is confronted with an unknown spectrum consisting of lines, bands, and noise from which it is hoped that accurate spectroscopic line parameters can be determined. Only then can meaningful atomic and molecular structures be analyzed and parameters calculated.

The process of line finding is an intermediate stage in the spectroscopic investigation of a light source, and in large part determines the ultimate accuracy of the structural parameters. In practice many hours are invested in preparation and observation of the source, and the analysis of the data may carry on for days, months, and even years. Yet more often than not the line finding, the process of translating a spectrum into line parameters, is performed rapidly on a computer without paying attention to or utilizing the subtleties of the spectral data. This imbalance of effort is reflected directly in the final data, since it is often the line finding that determines the error bars that characterize the accuracy of the published results.

In line finding, the parameters to be determined are wavenumbers, wavelengths, line strengths, and line shapes characteristic of the source, with accuracies limited only by the signal-to-noise ratio and photon statistics of the light. In this process the experimenter must correct the quasi-continuous data for the limitations and faults of the spectrometer, translate the data into a set of discrete spectral lines, and then calculate the spectral line parameters. Taken all together the line parameters constitute a line list. It is from this list that atomic and molecular structure and dynamics are determined.

To achieve high-quality results, it is evident that much attention must be paid to the mechanical, optical, and electronic design of the spectrometer and the conditions of observation, as well as the algorithms used to process the interferogram. Equally important are the algorithms used for line finding and line fitting. We will discuss in some detail here the theory and practice of determining spectral line parameters from a properly transformed and corrected interferogram. The discussion relies on our experience with *GREMLIN* and its predecessor, *DECOMP*, the spectral data analysis codes developed by Brault.

9.2 Line Finding and Line Shapes

The best way to prepare a line list is to choose an analytical model representative of the expected line shapes and then to match the model to the features in the spectrum by means of a least-square fit. This process can be time consuming. Often only line positions and rough intensities are needed, and simpler means for picking lines are sufficient. We will describe these simpler techniques first under the heading of *derivative line finding*. In practice, line positions are always picked by a derivative method, and when more accurate parameters are needed the lines are subsequently fitted by a least-square method.

9.2.1 Emission Lines

First, however, we need to say something about line shapes in general and about what is to be expected in the course of examining spectra from the usual light sources. Most emission line profiles are mixtures of gaussian and lorentzian shapes convoluted with the instrument function. The lorentzian part comes from the finite lifetimes of the energy levels and from pressure broadening, while the gaussian part comes from the doppler effect for radiating atoms in motion. The convolution of these two profiles is called a voigtian profile, which is expressible not as an analytic function but only as an integral. From line shapes and line positions we can deduce the desired atomic and molecular parameters and source operating conditions.

9.2.2 Absorption Lines

Absorption line profiles are different. The absorbance profile is voigtian in shape, just as for emission, but the observed *transmittance* profile is the *exponential* of this shape, due to the Lambert–Beer law of linear absorption. Furthermore, as the amount of absorption grows toward 100%, the lines become broad and flat-bottomed. It is the parameters of the absorbance that are important in determining atomic and molecular properties, not the shape or profile of an observed absorption line. Taking the logarithm of the observed spectrum produces an absorbance spectrum with line shapes that are only approximations to voigtian profiles. Lines can be picked as if they were emission lines, and even fitted, although the parameters lose much of their meaning unless the absorption is weak and the lines well resolved. The resulting line positions can be quite precise when the absorption is small, but the line shape parameters can only be considered qualitative.

9.3. Derivative Line Finding

The identification of features in the spectrum as spectral lines, called line finding or line picking, can be achieved by determining the zero crossings of the derivative of the spectrum. For a continuous function this is easy and accurate, but for a sampled spectrum containing noise the procedure is not uniformly accurate, because computing the derivative amplifies the noise and noise has a first-order effect at the crossover point.

An equivalent process for line picking is to convolve the spectrum with the derivative of a kernel or test function of some sort. We actually do this mathematically by maximizing the cross-correlation between a line in the spectrum and the chosen kernel. When the kernel function has the same shape or profile as the spectrum line in question, then we have an estimate of the the profile of the energy distribution about the line center, rather than just the position of the center. Of course, exact signal measurement requires least-square fitting to obtain statistically rigorous line shape parameters. But with a certain amount of *a priori* information about the line shape, it is possible to determine reasonably accurate line parameters by cross-correlation of the observed data with the expected line profile. These parameters are traditionally the intensity and width. With least-square fitting, the voigtian profile parameters are obtained, and then the line is completely characterized by intensity, width, and damping. The damping factor (lorentzian width/total width) is zero for a gaussian line shape and 1 for a lorentzian shape and is a measure of where the voigtian profile lies between its two extremes.

9.3.1 Kernel Functions

The best kernel functions to use are those that closely approximate the observed line shapes, such as a voigtian profile. But for convenience and easier computation, many simpler functions are in use that do not match any specific line shape. Some useful ones are illustrated in Fig. 9.1.

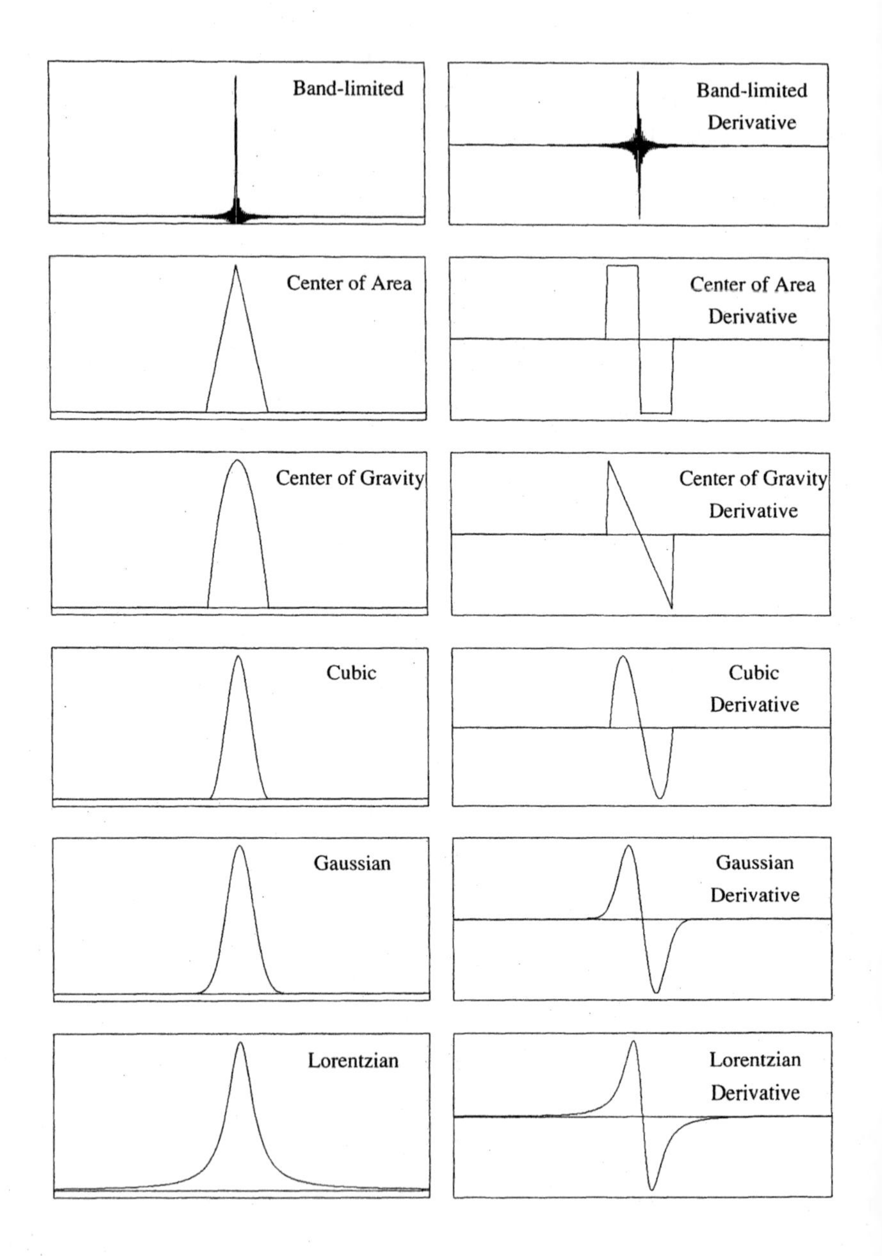

Fig. 9.1 Typical kernels and their derivatives for line finding algorithms: Band limited (sinc function); center of area; center of gravity; cubic; gaussian; and lorentzian.

There are several techniques for how these kernels are actually used and adjusted. One of them is to adjust some parameter of the kernel after examining the spectrum, the width, for example, and to keep it fixed for every line. Another is to use a kernel that can be adjusted during the line finding process. Least-square profile fitting, characteristic of this second method, lends itself to iterative refinement.

Let's look at some kernels to gain a perspective on their value and limitations. The kernel function acts as a *weighting* function. The simplistic approach to line finding is to take the numerical derivative of a spectrum and use the zero crossings of the derivatives to find the lines. Numerical differentiation is equivalent to convolution with the derivative of a sinc function that is truncated at some appropriate bandwidth (band limited). Such a kernel function and its derivative are shown in Fig. 9.1. The noise characteristics of this function are terrible, meaning that the noise is amplified relative to the original amplitude, and because of its complex structure many false lines may be introduced into the spectrum. It should never be used.

The-center-of-area kernel and its derivative, displayed in Fig. 9.1, effectively sample the spectrum on either side of the line and adjust the window until the two areas are equal. The abrupt truncation of the window distorts the results unless the width is carefully chosen to match the profiles (width and shape) of the lines in a particular spectrum.

Measuring the center of gravity is a more common technique, which is equivalent to convolution with a triangular function, Fig. 9.1. In practice the kernel is truncated, and the position of the truncation is an essential parameter in the noise sensitivity of this particular function. In comparison with the ideal cases of the voigt function at its gaussian and lorentzian limits, illustrated in Fig. 9.1, the center of gravity kernel samples the middle of the ideal kernel, which is equivalent to putting a parabola through the peak of the line.

The derivative functions for the center-of-gravity and center-of-area kernels are used with restricted ranges in order to reduce the noise sensitivity and systematic effects of baseline offsets. The abrupt truncation is vital to a practical implementation, but it also flaws the results.

A simple cubic and its derivative, shown in Fig. 9.1, with its width well matched to the lines under observation, provide the closest match to the profiles of the ideal kernels, have a smooth truncation, and give accurate results despite their relative simplicity.

For atomic and molecular spectra the ideal kernel is a voigtian function with a damping coefficient (lorentzian width/total width) appropriate for the spectrum. The limiting cases of pure gaussian and pure lorentzian kernels, illustrated in Fig. 9.1, suffice to illustrate the range of cases and show by comparison the limitations of the other methods. Ideally they should emphasize the portions of the line profile where the position information is largest. In contrast to the center-of-gravity and center-of area-kernels, the voigtian kernels are self-apodizing, which eliminates the truncation error and reduces the magnification of the noise in the convoluted spectrum. It is usually sufficient to choose a kernel that is dominantly gaussian with a modest amount of damping. For example, in our line finding routines we use a default damping of 0.1, where 0.0 represents a gaussian and 1.0 represents a lorentzian line shape. The critical criterion for determining the accuracy of the derivative line finding methods is matching of the kernel function line shape to the observed line shape. When the damping of the kernel function matches that of the observed line, the error is minimized and approaches that obtained from least-square fitting, in which all parameters are allowed to vary with each iteration of the fitting process. Usually a spectrum line is a function of four parameters: the line position σ, the width W, the amplitude A, and the damping ratio D.

9.3.2 Line Position Accuracy

The line position error introduced by the signal-processing algorithm in derivative methods reflects the inherent limitations of the data and the discrepancy between the expected continuous signal profile and the observed data. The data are discrete points, each separated by a reciprocal dispersion (cm^{-1}/point) from its neighbor. A spectral line profile has a characteristic full width at half maximum (FWHM) of N_w points. The vertical height of the peak is the signal strength S and has an associated uncertainty determined by the noise in the spectrum.

The discrete nature of the data directly affects the magnitude of the error. There may not be a data point at the peak of the spectrum line. The greater the number of data points across the line profile, the smaller the difference between the discrete representation of the profile and the actual profile. The uncertainty in the line position $\Delta\sigma$ due to discreteness is proportional to the square root of the number of points N_w in the width W and is also dependent on the line shape. Whatever the line shape, the relationship can be approximated by

$$\Delta\sigma \sim \frac{W}{\sqrt{N_w}}. \tag{9.1}$$

Now, how do we account for the effects of additive noise on our line parameters? A detailed analysis is lengthy and complex, but the result is that the error in position is the product of the factor representing the number of sampling points in a half-width, the signal-to-noise ratio S/N, and a geometrical factor f relating to how well the kernel function matches the line shape:

$$\Delta\sigma \sim \frac{W}{\sqrt{N_w}} \cdot \frac{f}{(S/N)}. \tag{9.2}$$

A useful rule of thumb is: The half-width of the line can be divided into a number of parts equal to the signal-to-noise ratio, and the line position error is no larger than one of these parts. To illustrate, for a line with four sampling points in a width of 20 mK and $S/N = 10^4$, $\Delta\sigma = 10^{-3}$ mK, as is typical in the infrared. In the ultraviolet, for a typical line width of 200 mK and a signal-to-noise ratio of 10, the error is 10 mK.

Can we now relate the shape of the kernel function to the relative error in line position due simply to the choice of convolving function? As Fig. 9.1 demonstrates, there is considerable variation in the degree of resemblance between the functions in the top two rows and the ideal gaussian and lorentzian functions in the bottom two rows. If used with great care, the center-of-area and center-of-gravity functions are better than they appear. When optimum ranges are used (not always a simple thing to do for spectra of mixed widths), the error for a restricted center-of-area kernel is larger than for a kernel that has the exact shape of the line, while that for the restricted center-of-gravity kernel is still larger. Of the simple functions, the cubic combines a smoother truncation and less noise than either of the other functions and produces good results if the width is chosen carefully.

9.3.3 Useful Line-Finding Properties of First and Second Derivatives

Some useful properties of derivative convolution methods for line finding are the following:

1. *Suppression of broadband features.* The distinction between lines and broadband features is determined simply by the width of the kernel function, and choosing this width appropriately focuses the spectral analysis on the desired features in the spectrum.

2. *Insensitivity to baseline shifts.* Baseline sensitivity is eliminated to first order because differentiation removes quasi-constant features from the spectrum,

and the estimation of the line parameters is independent of the location of the zero. In particular, the separation between inflection points is nominally the same for any background level.

3. *Consistent determination of parameters for all lines in a spectrum.* Once the kernel function is chosen, it is used for all lines in the spectrum. This produces a consistent determination of parameters, but it does limit to a single basic shape the types of lines that can be measured to a consistent precision with a single operation.

9.3.4 Techniques of Derivative Line-Finding

A line-finding scheme that combines numerical first *and* second derivatives of a voigtian kernel has several appealing features. For spectra with blended lines, a second-derivative, or inflection point, method can often separate overlapping lines. However, each differentiation process amplifies noise in the convoluted spectrum, and hence for unblended spectra a first-derivative method yields slightly more accurate line positions. Another option is to take the average of the first- and second- derivative positions when possible, to separate lines and retain the greater accuracy.

The method goes as follows. A numerical algorithm convolutes the spectrum with the derivative of a kernel and then seeks zeroes in the convoluted spectrum. Once a zero crossing is located, the second derivative function is stretched (interpolated) by a factor of 2 to improve the determination of the line position. The process is illustrated in Fig. 9.2. A spectrum of five lines with a signal-to-noise ratio of 1000 is illustrated in Fig. 9.2a, the first-derivative spectrum in Fig. 9.2b, and the second-derivative spectrum in Fig. 9.2c. Clearly the first-derivative spectrum is not suitable for measuring the separations of the blended lines. Two of them can be separated in the second-derivative spectrum.

9.3.5 Noise and Filtered Derivative Line-Finding

The technique of derivative line-finding can amplify the noise disproportionately unless the spectrum is filtered and the high-frequency noise spikes smoothed out. By way of illustration, review the sample spectrum and its derivatives shown in Fig. 9.2, and look at Fig. 9.3 for their transforms. The effect of taking derivatives is to amplify the high-frequency noise with respect to the peaks. A good practical approach to suppressing this noise is to filter the derivative kernels with a gaussian filter to suppress the sharp components. The measured line parameters can easily be corrected for the effect of the filter, which is to increase the line width and reduce the amplitude, while keeping the area constant. In this context, *to filter* means "to convolute with a (gaussian) filter function," or correspondingly to multiply the interferogram by the transform of the filter function. In another context in Chapter 6, filtering in connection with interpolation meant multiplying the interferogram with a filter function directly.

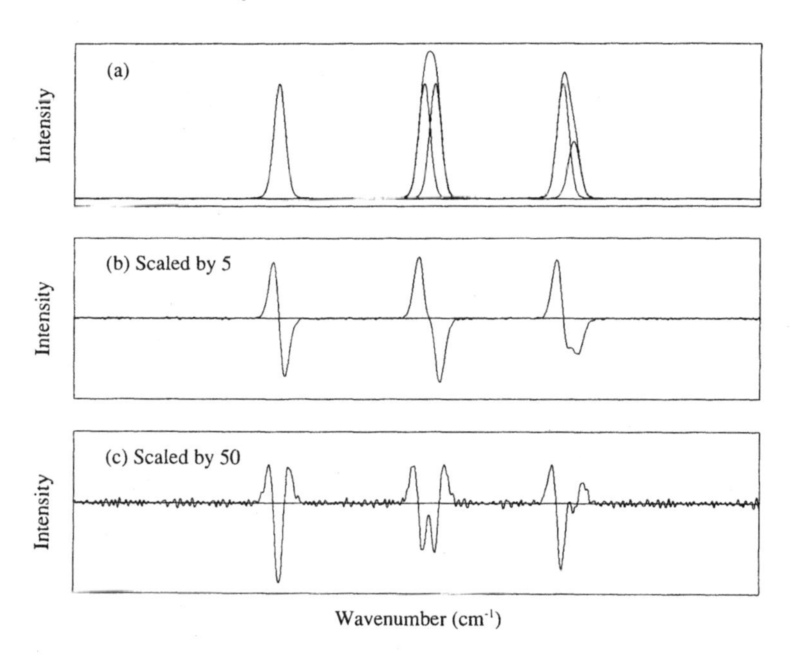

Fig. 9.2 First and second numerical derivatives of a spectrum with noise for the determination of line positions and for assessing the ultimate resolution of spectral lines. (a) Individual line components superimposed upon the raw spectrum. (b) First-derivative spectrum. (c) Second-derivative spectrum. Note that the first-derivative is able to resolve the asymmetric blend but not the symmetric doublet. The second-derivative resolves this doublet. The signal-to-noise ratio is degraded from 1000/1 in (a) to 10/1 in (c).

Another application of filtered derivative convolution is to suppress distortion of line detection by ringing, as mentioned earlier in connection with apodization. Many sources produce a spectrum with lines that are sharper than the resolution width of the spectrometer, and the lines exhibit ringing. The ringing consists of multiple sidelobes with appreciable intensities, and consequently all the line-finding schemes discussed earlier pick each sidelobe as a real spectrum line. In this manner, a spectrum of a few thousand lines becomes a line list of tens of thousands of lines, of which the majority are fictitious. One method of picking only the true spectrum lines is to filter the derivative kernel (not the spectrum) by convoluting it with a gaussian profile of sufficient width to suppress the ringing. Figure 9.4 shows two spectrum lines with ringing. An ordinary line finding technique as shown picks the ringing as well as the two real lines. Every line except the real ones can be eliminated by the proper choice of a gaussian filter. It should be noted that the lines have a shape far from a voigtian profile and that the resolution is degraded by the filter. A better method of treating ringing is discussed later, in Section 9.4.7.

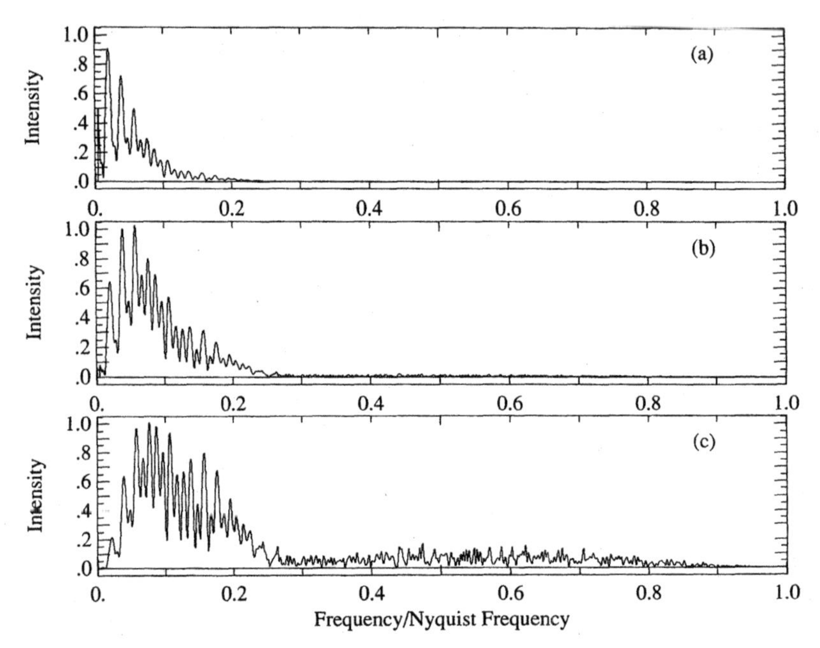

Fig. 9.3 Fourier transforms of the spectrum in Fig. 9.2: (a) Raw spectrum transform; oscillations are the result of the interference between the five lines. (b) First-derivative spectrum transform scaled by a factor of 100. (c) Second-derivative spectrum transform scaled by a factor of 500.

9.3.6 Resolution and Line Finding

There are several criteria for defining the resolution limit, which is the minimum resolvable wavenumber difference between two equally intense lines that can be detected with a spectrometer. The Rayleigh criterion and the Sparrow criterion are among the best known for grating spectrographs. Although not directly applicable to FTS spectra, they will illustrate the point we are making. The Rayleigh criterion for the resolution of two equal-intensity lines requires that the principal maximum of one line coincide with the first minimum of the adjacent line, producing a central intensity of 81%, as illustrated in Fig. 9.5. The Sparrow criterion requires only that the lines be sufficiently separated for the combined profile to have a zero second derivative halfway between the line centers.

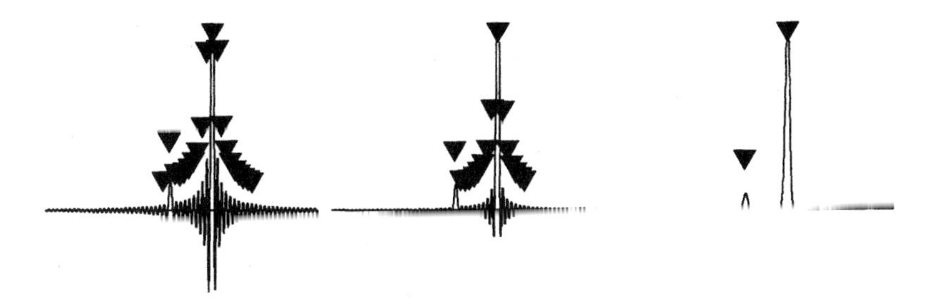

Fig. 9.4 Filtered line-finding. Unfiltered spectrum; two-point gaussian filtering; four-point gaussian filtering. The triangles indicate estimated line positions.

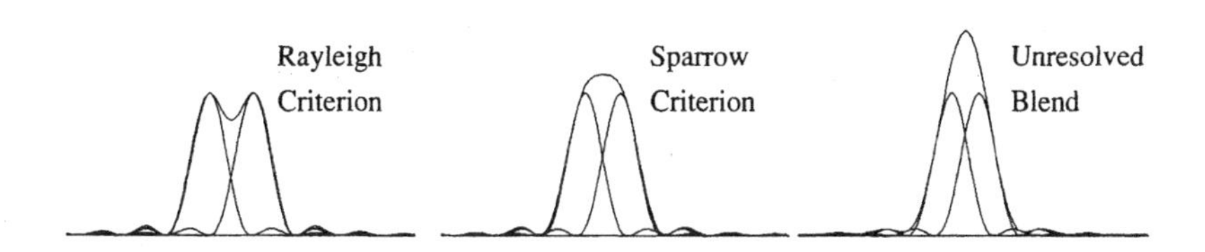

Fig. 9.5 Resolution criteria and the separation of blended lines: Rayleigh, Sparrow, and unresolved.

While the criteria are useful for visual interpretation of line profiles, they do not address the actual limit of detection that can be achieved with numerical signal-detection algorithms. In particular, if the observed line profiles are amply sampled (more than five samples per FWHM), then the number of successive differentiations necessary to distinguish the spectral features is a more quantitative definition of the criterion for resolving spectral lines. Figure 9.2, previously used when we were discussing differentiation and noise, demonstrates the process of resolving closely spaced lines.

The derivatives of the single line at the left of Fig. 9.2 are predictable, with one and two zero crossings for the two derivatives. The central pair of lines are separated by the Sparrow criterion where the top of what appears to be a single line is flattened and the overall width is larger than a single line. The small kink in the first derivative is sufficient for detection, but the second derivative is necessary for confirmation. The second pair is more easily resolved in both derivatives owing to the differing intensities of the doublet. As is evident, differentiation degrades the signal-to-noise ratio because the derivative kernel used the full bandwidth, whereas the spectrum is band limited and some filtering may be necessary. There is a trade-off between detectable resolution and noise, but loosely speaking the number of components that can be resolved in a blended line structure is equal to the number of derivatives that can be taken before the line becomes indistinguishable from the noise. This limitation does not apply to line shape fitting with a least-square method and a well-matched kernel profile.

9.4 Least-Square Fitting of Line Profiles to Voigtian Functions

The central issue of this section is the accurate measurement of line positions and shapes in the presence of noise. We can find the approximate positions of the lines by derivative methods, but now we want to focus on accurate *fitting* of line profiles by least-square techniques rather than by convolution methods. Again, we do so in the context of the *GREMLIN* data analysis code. We will take as a line shape model the voigtian profile, which is appropriate for modeling emission lines of atomic and molecular species and, in favorable cases, also for absorption spectra.

9.4.1 Voigtian Functions

The voigtian function is used in modeling emission line and absorption line profiles of atomic and molecular spectra in homogeneous pressure and temperature conditions, as, for example, in nonsaturated absorption in the atmosphere. The function is defined by the convolution of a gaussian and a lorentzian function. Experimentally this has a sound basis, because the gaussian line shape is characteristic of a doppler profile, a lorentzian is characteristic of natural and pressure-broadened line shapes, and spectral lines exhibit both these effects. The full width at half maximum T of the voigtian function can be thought of as being made up of two components, a lorentzian with FWHM of L, and a gaussian of FWHM of G. We characterize the mixing or proportion of each by the damping parameter $D = L/T$, where a value of 0.0 signifies a pure gaussian line shape and 1.0 signifies a pure lorentzian. A common notion is that the gaussian is a bell-shaped curve while a lorentzian is sharply peaked and has large wings. The fact is that at their peaks they are nearly indistinguishable. It is in their wings that they differ greatly. Figure 9.6 shows these shapes. In some contexts other authors define damping as the ratio of lorentzian to gaussian widths. The damping then spans values from zero to infinity, which makes numerical calculations awkward.

The narrowest line is characteristic of atomic or molecular lines from a low-temperature electric discharge in gas at a pressure of a few torr (mbar). The intermediate curve is characteristic of visible emission from a gas or vapor in a furnace at perhaps 1500 K and a pressure of 50 torr. The widest curve (lorentzian) represents emission from the same furnace at a pressure of several hundred torr or in the low-wavenumber infrared at a pressure of a few torr.

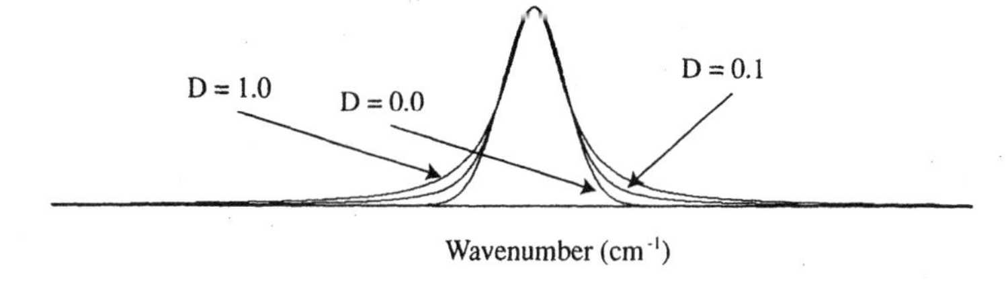

Fig. 9.6 Voigtian functions with $D = 0.0$ (gaussian), 0.1, and 1.0 (lorentzian) in an emission spectrum.

9.4.2 Fitting Considerations

The least-square fitting method allows us to fit a prescribed model to the data. It makes iterative comparisons of the model to a set of experimental data and readjusts the model parameters with each iteration until the sum of the squared deviations between the model and the data is a minimum. The model for emission lines is the voigtian function, as just described, and fitting algorithms are designed specifically for this function. If the line shapes under examination are not describable as voigtian functions, then the effectiveness of the fitting process is diminished significantly, and a different procedure or a new model must be used. In emission spectra, the most commonly encountered example of a nonvoigtian profile is the self-reversed spectrum line. As the optical depth of the source becomes large and the center of an emission line is eaten away by absorption, the line shape ceases to resemble a voigtian function, and a fit is not possible. We can still *pick* the line with one of the derivative kernels, but we cannot derive useful line shape parameters, as mentioned earlier. Similarly, strong absorption lines and laser, grating, and magnetic resonance spectra often display other profiles. Also, the ringing that accompanies underresolved spectrum lines causes the central line shape to depart significantly from a voigtian profile, aside from the extraneous lines, and complicates the fitting process. Nevertheless, because the voigtian function is sharply peaked and has four parameters — the wavenumber σ, amplitude A, width W, and damping D — it can be used to fit many line shapes even though the significance of the parameters is partially lost. The fitting process is nonlinear in σ, W, and D, and therefore may require many iterations and adjustments of parameters until a best fit is obtained.

An example may put this in perspective. Consider a four-line spectrum consisting of a gaussian line, a lorentzian line, and a blended doublet made up of two overlapping lines, shown in Fig. 9.7a, with a signal-to-noise ratio of 100 (peak to r.m.s.). Since the doublet is asymmetric, a simple line finding identifies four lines, as in Fig. 9.7b. The line finder assumes a minimal damping ratio D, and as a result the gaussian has the smallest residual, the lorentzian a larger residual, and the doublet the largest residual, due to the uncertainty in the relative intensities of the components. After five iterations of least-square fitting, shown in Fig. 9.7c, the single lines are well fit, as indicated by residuals that are insignificantly different from the adjacent noise. However, the doublet has significant residuals that are antisymmetric, indicating that the fitting algorithm is having difficulty resolving the separation between the components of the doublet. Ten additional iterations are sufficient to permit the algorithm to locate and fit the amplitudes and shapes of the doublet components, as indicated in Fig. 9.7d.

Qualitatively, achieving a residual spectrum that contains only uniform (white) noise is an initial indication of the adequacy of the fitting process. However, whenever the lines are overlapping, which is most of the time, considerable judgment must go into the proper selection of line positions and monitoring of the fitting process. For example, if the spectral lines are all produced by the same source, at a given temperature and pressure, then we can constrain the widths and damping ratios of each line to a value determined for the strongest, unblended lines. In such a fashion, it should be possible to fit all the lines in a spectrum and establish a degree of confidence in the line parameters. Numerically, the r.m.s. residuals within five half-widths of the line centers are translatable into a measurement error. Equation (9.2) suggests that a well-sampled line (four samples in the FWHM) should be measurable to 1 part in 200.

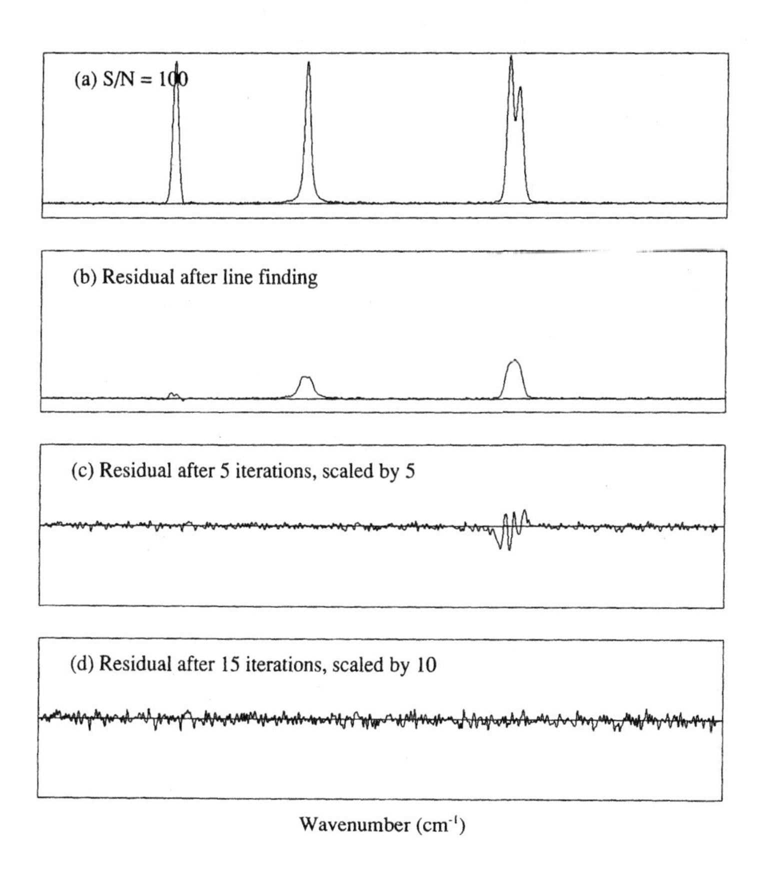

Fig. 9.7 Least-square fitting of spectra to voigtian functions: (a) Four-line spectrum containing a blended feature, with a signal-to-noise ratio of 100:1 peak to r.m.s. b) Residual spectrum after line finding. (c) Residual spectrum after five iterations of fitting. Note that the single lines are easily fitted, whereas the blend requires more iterations properly to position and fit the component lines. (d) Residual spectrum after 15 iterations, with residuals containing no systematic features.

9.4.3 Special Problems in Fitting Voigtian Functions

The four first-order partial derivatives of the voigtian function with respect to its four parameters provide some insight into the hazards of numerical analysis and the utility of being able to examine functions in both Fourier domains.

The *amplitude* derivative is simply the line shape itself. The positional derivative with respect to *wavenumber*, the familiar one, is asymmetrical. The derivative with respect to the *width* is more complicated. It has two maxima and a zero in the center. The derivative with respect to the *damping* is even more complicated. As the spectrum line gets narrower, the transforms of the derivatives extend to higher frequencies, and some interesting consequences arise. These derivatives and their transforms are shown in Fig. 9.8.

The transform of the positional derivative has a small nonzero value at the Nyquist frequency and consequently introduces a small amount of aliasing. The width and damping derivatives have even larger values at the Nyquist frequency and introduce still stronger aliasing. To counteract this problem, we must increase the number of samples. For a gaussian profile it is sufficient to *observe* data that will produce three samples in one full line width in the spectrum, so far as the instrument is concerned, but when we *fit* the spectral data it behooves us to change our style a little and increase the number of samples by a factor of 2 at least. A lorentzian requires 10 samples in a line width in order to be fitted properly!

We can increase the number of observed spectrum samples by zero-filling the interferogram or by Fourier interpolating the transformed interferogram, called *post-transform stretching*, as discussed earlier in another context. Zero-filling is mathematically equivalent to convolving the data with an infinite sinc function, which can be performed after the data have initially been transformed in the normal fashion. The process of zero-filling should be distinguished from sampling more often, which changes the free spectral range. We want to sample more frequently to obtain more samples across a line profile, rather than to increase the resolution beyond what is necessary, which simply pours noise into the data.

9.4.4 Fitting Errors from Sampling

A study of parameter-fitting errors for voigtian profiles has produced some rules of thumb for estimating accuracies. Line positions can be measured with high accuracy with only two to three points in the full width at half maximum, while line shape parameters can be measured to at least 1% for eight points in the full width or better than 0.2% for 16 or more points in the full width.

9.4.5 Fitting Errors from Noise

The presence of noise complicates the estimate of parameter errors, but a good working expression is

$$\frac{\Delta\sigma}{W} \sim \frac{\Delta A}{A} \sim \frac{\Delta W}{W} \sim \Delta D \sim \frac{\Delta E}{E} = \frac{k_p}{\sqrt{N_w}} \cdot \frac{1}{(S/N)}, \tag{9.3}$$

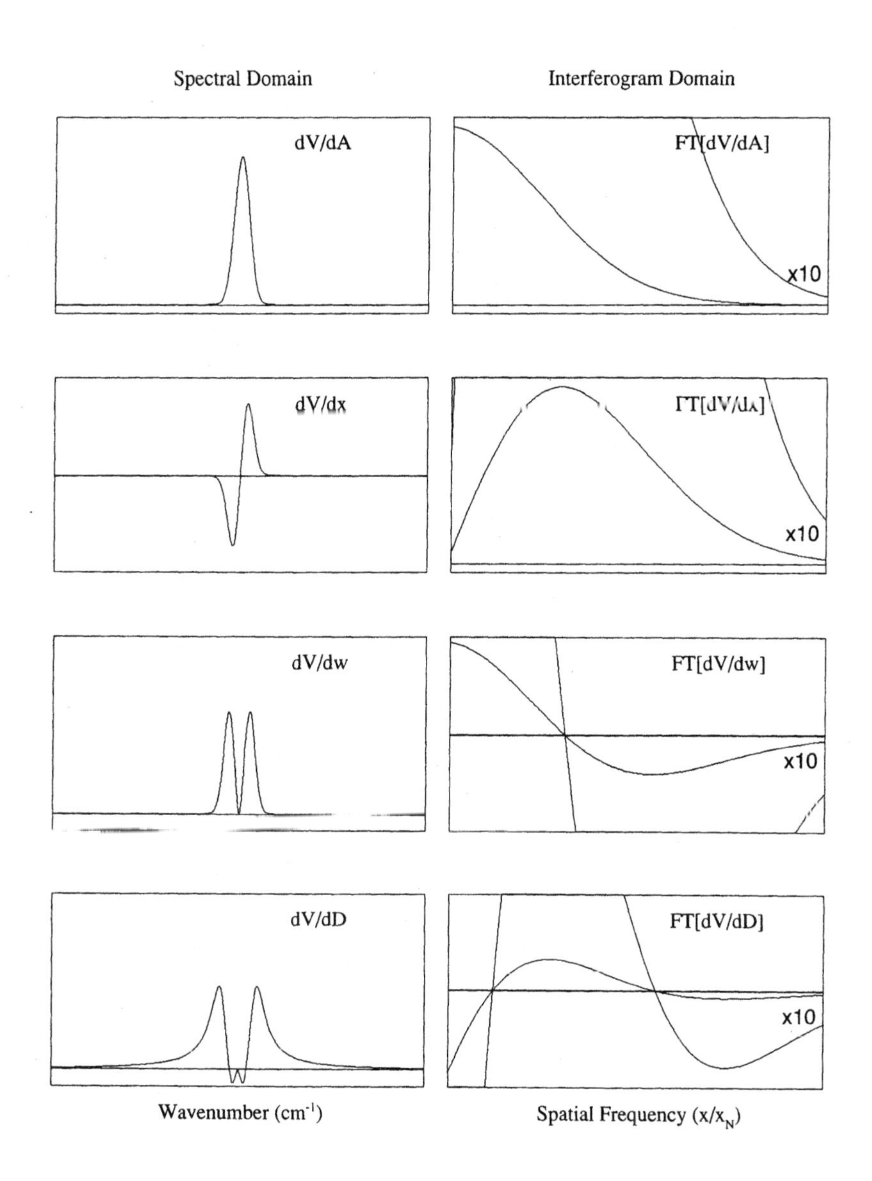

Fig. 9.8 Derivatives of voigtian functions and their transforms.

where σ is the wavenumber, A is the line peak amplitude, W is the line width, D is the damping (shape parameter), E is the area under the line, S/N is the signal-to-noise ratio, and N_w is the number of statistically independent points in a line width. k_p, an empirical weighting factor, has the approximate values

$$k_\sigma = 1;\ k_A = 1;\ k_W = 2;\ k_D = 4;\ k_E = 2.$$

9.4.6 Fitting Errors from the Sampling Grid

The sampling grid used by the computational algorithm cannot sample each line at the optimum position. Under high dispersion the lines often appear asymmetrical with a tilted flat top, simply because a grid line is not at the exact peak wavenumber. When a line peak falls exactly on the grid or exactly in between two grid lines, the line shape is most accurately fitted, except for the flat top at the halfway position. In other cases the line parameters show systematic errors that are greatest when lines fall at the one-quarter or three-quarter points on the grid. The errors in shape and amplitude are small but easily detectable for lines of moderate S/N. In all cases, zero-filling makes these sampling grid errors insignificant. Just don't forget to do it.

9.4.7 Fitting Underresolved Spectra

The fitting processes just described assume that the data were sampled with sufficient resolution to permit satisfactory fitting using voigtian function derivatives. Often this is not the case, and when the number of samples per line width falls below 3 to 5 points, the observed line shape is not a voigtian but is dominated by the instrument profile and shows ringing or sinc function sidelobes.

Underresolved spectra occur when the instrumental line width is equal to or larger than the spectrum line width. Equivalent ways of saying this are: the resolution of the interferometer is less than the spectral line width; the scanning length of the interferometer is too short; the interferogram was cut off before its amplitude reached zero; the interferogram is band limited. The problem is most noticeable in the infrared, where doppler widths are small.

The ringing can be suppressed by filtering the spectrum, but the line shape is seriously degraded from its true shape and the resolution is reduced. The results of convoluting a gaussian filter with the spectrum, which is equivalent to multiplying the interferogram by a gaussian, is illustrated in Fig. 9.9. The true line with unlimited resolution is shown in part (a), together with the envelope of the interferogram that produced it. In part (b) the interferometer scan was stopped before the interferogram amplitude reached zero. The resulting spectrum line shape is shown, together with the amplitude of the interferogram, which is now truncated and consequently band limited. The ringing is pronounced. In Fig. 9.9 (c) a gaussian filter multiplying the truncated interferogram produces the same band limited interferogram but with an amplitude envelope that has a smaller discontinuity at the band limit. The spectrum line is widened, the ringing has decreased in amplitude. In Fig. 9.9 (d) a wider filter has been applied, and there is no longer a discontinuity in amplitude of the interferogram at its extreme. The spectral line now shows no ringing, but has a smaller amplitude and larger width than the true line shape in part (a).

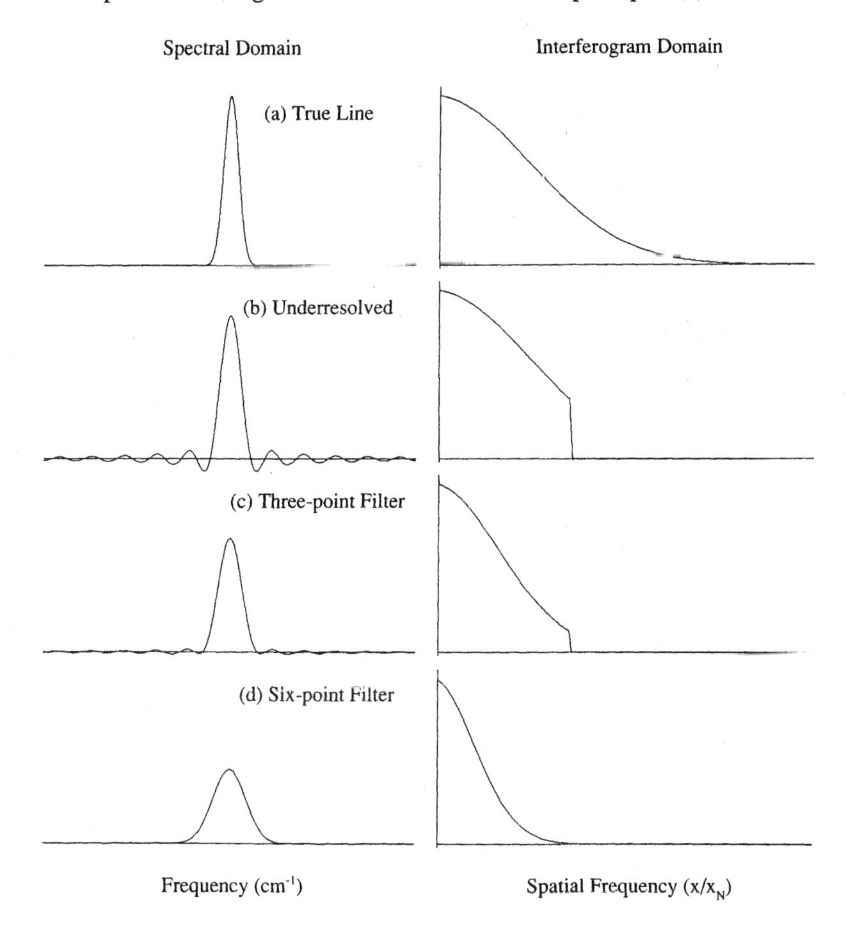

Fig. 9.9 The use of spectral filtering to remove the instrument function (ringing). (a) True line profile and its transform. (b) Underresolved line with ringing. (c) Three-point gaussian filter convoluted with the spectrum. (d) Six-point gaussian filtering. The ringing is suppressed, but the line shape is degraded.

There is a method of voigtian profile fitting that removes the ringing without modifying the line shape or resolution. When a spectral line is fully resolved and the interferogram envelope has no discontinuities, a gaussian profile for picking and fitting the line is used. When the line shows ringing, a modified voigtian profile that shows the same ringing as the spectral line is used. The interferogram is cut off at some point N. We take the voigtian profile, get its transform (its interferogram), truncate this interferogram at N, transform it back to spectral space, and use this band-limited voigtian profile as the fitting function. With each least-square iteration, we are fitting the ringing as well as the central maximum with the same shape as the line itself. In the final fit, the ringing residual has been reduced to insignificance. What follows is a detailed discussion of this method.

It is a fact that most of the line shape information in a spectrum is contained in the small path difference portion of the interferogram near the origin. Consequently, voigtian fitting with a restricted bandwidth corresponding to the bandwidth of the measured interferogram allows for line shape determination from an incomplete interferogram. This method convolves the model spectrum with the instrument function during the fitting process, producing a spectrum containing the fitted lines and the noise without any distortion.

In the Fourier (interferogram) domain, the fitting algorithm looks at the total residual and consequently fails to fit the spectrum line accurately. The fitted line profile underestimates at delays (path lengths) smaller than the cutoff frequency (maximum path length) so as to be able to overestimate the residuals for the delays larger than the cutoff frequency. Such a result yields the minimum rms residual, but is a poor representation of the data because the algorithm is looking at the full bandwidth when there is no information at delays beyond the cutoff.

In order to obtain sensible results, the algorithm must be constrained so that it ignores all Fourier components beyond the cutoff delay. To do this, after each iteration of fitting, the residuals are trimmed with a sharp cutoff filter matched to the correct delay. After the first iteration of filtered fitting, the Fourier components accurately represent the small delay components and the area between the profiles shows the convergence of the residuals toward the underresolved profile. If the assumption is correct that the voigtian model is an accurate one for the line shape, then the parameters determined by the process of filtered fitting are the correct ones, subject to the same limitations as normal voigtian fitting. The saying is, "If you can model the instrumental function, you can remove its effects." Figure 9.10 illustrates the inability of a traditional fitting process to properly fit underresolved spectra. In contrast, with ring fitting the algorithm converges and properly models and removes the instrumental line shape from the spectrum (Fig. 9.11). The algorithm is demonstrated on an underresolved spectrum of OH (Fig. 9.12) with a dynamic range of 5000: The strong lines are properly fit and the instrumental function properly modeled without distorting the adjacent weak lines.

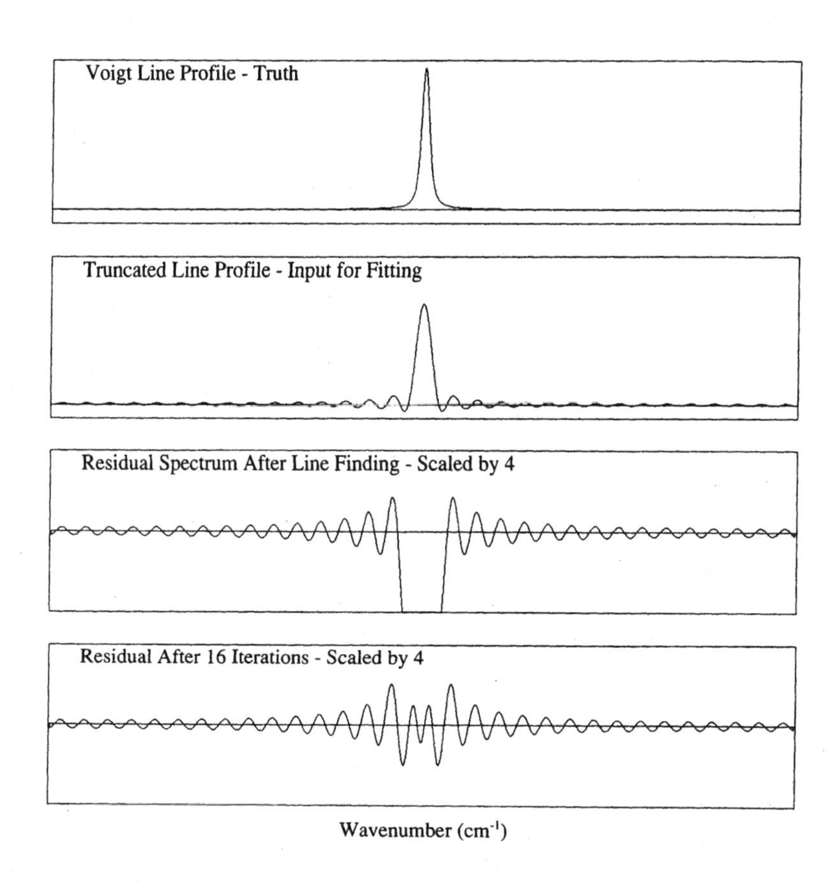

Fig. 9.10 Traditional least-square fitting (without ring fitting). The errors are on the order of 15%.

9.4.8 Rates of Convergence

For isolated lines that are unblended, a linearized iterative least-square fit converges toward a solution quickly, within about 10 iterations for a signal-to-noise ratio of 100, unless the ringing is very strong, as in Fig. 9.10 For blended or unresolved lines, more are necessary. With unblended lines, the shapes are typically symmetric and the line position converges first, since the positional derivative is the only odd function of the voigtian derivatives. The three even derivatives compete for the residuals beneath the line and converge more slowly. The most difficult case for a fitting algorithm to handle is where adjacent lines interact, because the algorithm adjusts the residuals within five half-widths on either side of the line center. Unless great care is taken with choosing the initial

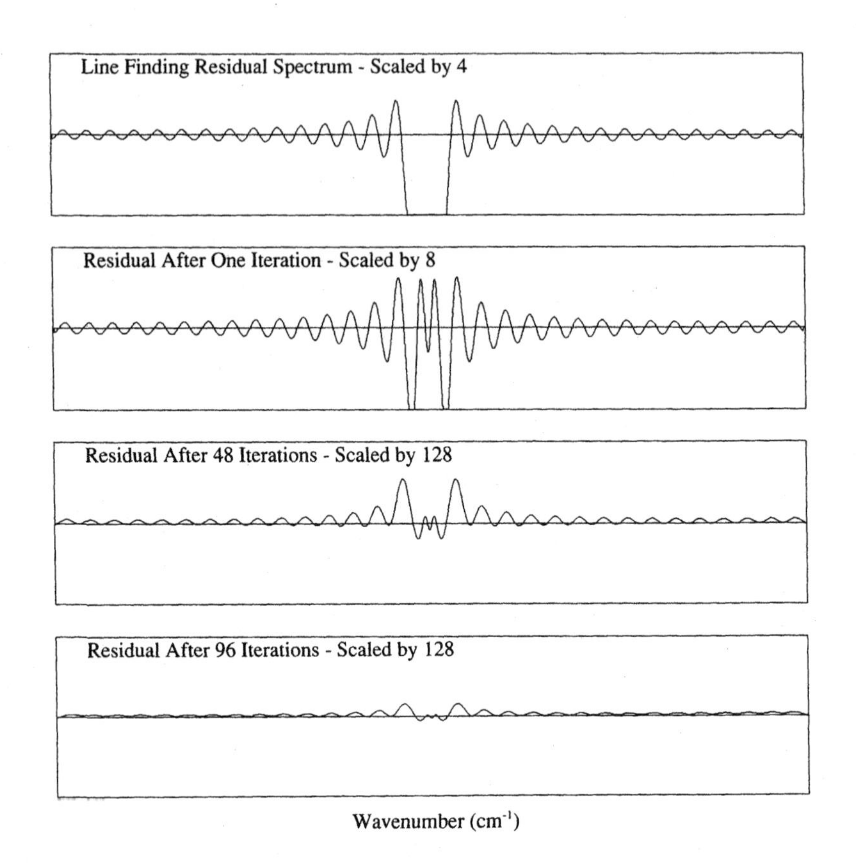

Fig. 9.11 Least-square fitting with ring fitting. The errors are on the order of 0.01%.

parameters of position, intensity, and width, it is possible to get false fits, with some lines narrower and more intense and others wider and less intense than they were originally. To prevent this result, the width and damping can be set to fixed values consistent with the expected line shapes and widths measured in unblended regions of the spectrum. An example of the algorithm in action (Fig. 9.12) illustrates the complexity and necessity of filtered fitting of underresolved data. The raw spectrum has a dynamic range of 5000 [the plot is a linear (1–10) – log (10–10 000) plot that allows the weak features ($S/N \sim 5$) to be visible on the same scale as the strong lines]. The sidelobes of the instrument function are approximately 100 times stronger than the weak lines and massively distort the spectrum. After fitting, the strong lines are properly represented without instrumental distortion, and the weak lines (including several previously hidden by the instrumental function) are similarly well represented.

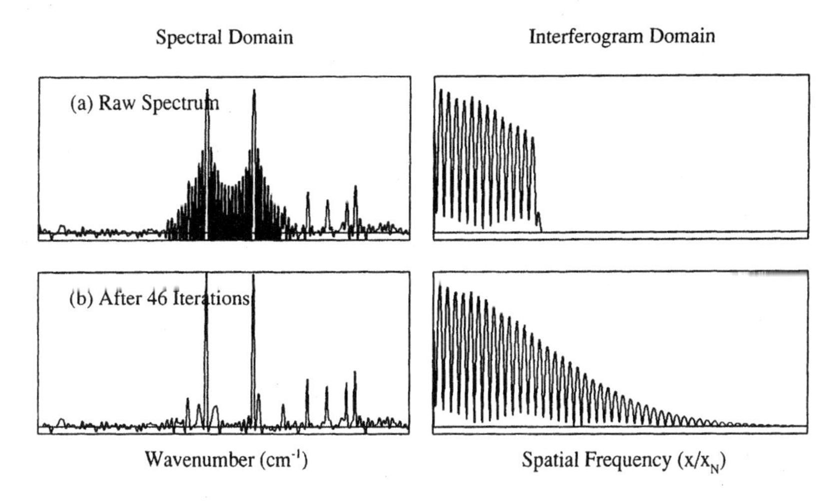

Fig. 9.12 Line fitting with filtered residuals (ring fitting). (a) Observed spectrum with ringing owing to inadequate resolution plotted on a linear-log scale used to accentuate weak lines. (b) Reconstructed spectrum after 46 iterations.

9.5 Areas and Equivalent Widths

In astronomy, where spectral lines are primarily absorption lines, the equivalent width of a spectral line is the width of a rectangular profile that extends from the continuum on either side of the real spectral line to zero absorbance and that contains the same area as the real spectral line (Fig. 9.13). This width is a measure of the energy removed from the continuum by absorption. It is a useful parameter of a spectral line, computed in our algorithm at the same time the line is fitted.

The actual numerical value of the width, which represents an area, depends on the dispersion with which the spectrum is plotted and on the height of the continuum. The fractional amount of absorption is independent of the dispersion and absorption. The dispersion is constant and need not be scaled or corrected for when comparing equivalent widths in the same spectrum, but the background does need correction. What's important is the relative area — relative to the background continuum. So to compare spectrum lines, the background continuum must be normalized to the same level throughout the spectrum by a ratioing process. Absorption lines — say, of 10% depth — must still be at depths of 10% when the background is corrected, no matter what the original background was at each line. A common practice is to set the continuum to 1.0 and complete absorption to 0.0, by ratioing a smoothed continuum to the observed spectrum. Then the absorption spectrum is changed to an absorbance spectrum by taking the natural logarithm of the data points, the lines are fitted, and the equivalent widths are determined.

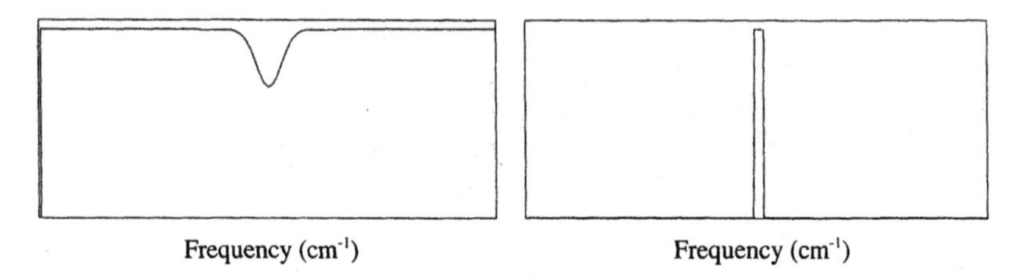

Fig. 9.13 Equivalent width (area) of a spectral line in absorption.

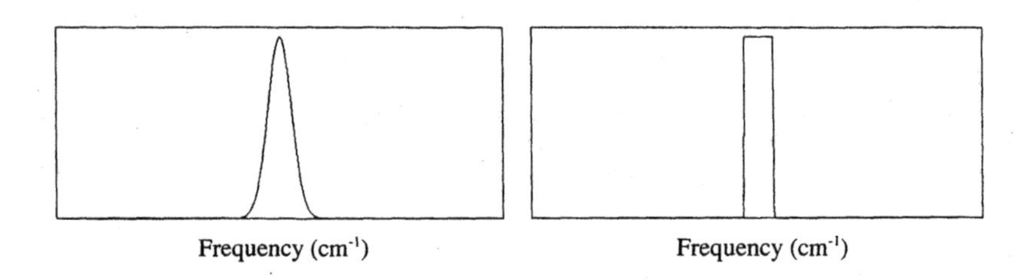

Fig. 9.14 Equivalent width (area) of a spectral line in emission.

The equivalent width of an emission line, for which there is assumed to be no continuum, is defined as the width of a rectangular profile that has the same central intensity and the same area as the line (Fig. 9.14). The equivalent width must be coupled with the intensity to get the energy emitted. Individual authors often give exotic definitions for equivalent widths, yet all are understandable when you keep in mind what the purpose is. For example: "The equivalent width for H_α emission superimposed on a continuum produced by free - free emission in a region of ionized hydrogen is defined as the number of angstroms of continuum necessary to give the same luminosity as the H_α line."

9.6 Wavenumber Calibration

Once the lines have been picked and fitted, a wavenumber scale needs to be established and then calibrated for absolute accuracy. The scale is determined by two quantities: the distance in vacuum between samples, δx, and the cosine of the angle θ between the optic axis and the direction of the light being measured, as discussed in Section 5.2. A preliminary scale is established by using the control laser wavenumber for distance determination, which is accurate enough for many purposes. In the usual reduction of the interferograms, we do not take into account the finite aperture effect on the wavenumber scale and usually assume that $\theta = 0$ for both the measuring laser and the main beam (no off-axis light). This assumption is only approximately true, since the aperture is never vanishingly small, and the amount of off-axis light depends on the state of alignment of the interferometer as well. However, theory is useful in telling us that virtually all known nonideal effects result in a simple stretching or compression of the whole wavenumber scale, as shown in Eq. (5.12), so

$$\sigma_{\text{true}} = (1 + k_{\text{eff}}) \cdot \sigma_{\text{apparent}} \tag{9.4}$$

or

$$\lambda_{\text{true}} = (1 + k'_{\text{eff}}) \cdot \lambda_{\text{apparent}}. \tag{9.5}$$

If we have a single spectrum line whose absolute wavenumber is known, we can use this one line to set the correction factor absolutely, without needing to have many standard lines distributed throughout the spectrum.

In some instances, the control laser line itself may be observed in the spectrum due to scattering at the cat's-eye secondary mirror or elsewhere, so we might consider using the laser wavelength to find the correction factor. We still wouldn't have an absolute wavenumber calibration, because the effective angle for this scattered light is unknown, and in fact the exact wavenumbers of the lines emitted by typical commercial lasers are unknown. In principle in a He-Ne laser they can vary by a significant fraction of the doppler width of the neon line, depending on the exact characteristics of the laser stabilization circuit used.

In practice, k_{eff} is best determined from a gas *within the source* that emits or absorbs a standard wavelength. We often observe correction factors considerably larger than that expected for the finite aperture ($\sim 5 \times 10^{-7}$) up to 3×10^{-6}. The correction factor is normally negative; raw wavenumbers must be decreased or wavelengths increased, because of the cosine factor mentioned earlier.

In the laboratory, lines of N_2O in an absorption cell and argon in emission are the most useful for absolute wavenumber calibration, even though emission lines are subject to pressure shifts. N_2O lines are accurate to 0.000 01 cm^{-1} for the band at 4400 cm^{-1}. Argon wavenumbers are accurate to 0.0001 cm^{-1} in the infrared. For astronomical sources the N_2O cell can be used; or, alternatively, atmospheric O_2 lines are useful. Wavenumbers of these lines in the (γ)(2-0) band near 6300 angstroms are known to a relative accuracy of 0.000 26 cm^{-1} and are valid for astronomical observations when the local wind speed is corrected for and the air mass is less than 3. However, these lines are subject to pressure shifts, and their shapes are noticeably asymmetric. Atmospheric water vapor lines should be avoided as standards because they are subject to even larger pressure shifts, which are unpredictable and variable, owing to a lack of thorough mixing in the air.

10

PROCESSING OF SPECTRAL DATA

Once we have collected spectral data, we need to reduce them to numbers and plots that are useful to astronomers, physicists, chemists, and scientists in general. Atlas plots of the spectrum of the radiation at both low and high dispersions are important. The former gives an overview of the intensity distribution, and the latter shows details such as line shapes and widths. Sometimes the data can simply be plotted without any modification or analysis. However, the spectral data we get from an instrument are not a true representation of the radiation coming in; rather, they are that radiation as modified by the instrument. As we have discussed, the spectrometer output for purely monochromatic radiation input is not an infinitesimally narrow spectrum "line," because of the finite aperture and resolving power of the spectrometer. We want the final atlas to show as closely as possible the true spectrum of the radiation itself, with instrumental effects removed as far as possible. Of course, presenting the raw spectrum itself is desirable as well, in order to show what data the experimenter has to work with.

Another need is for a line list, which includes all lines evident on the high-resolution atlas and which incorporates accurate fitted line parameters, including frequency and wavelength, intensity, width, area, and damping, or an equivalent set of line parameters.

In this chapter, we shall discuss how to make atlases and line lists. All of our examples are drawn from data taken on the McMath–Pierce Fourier transform spectrometer at the National Solar Observatory, Kitt Peak. The data-processing code is *GREMLIN*, the successor to many versions of *DECOMP* authored by J. W.

Brault over the last 25 years. There are other codes extant for similar data reduction that are adaptable to everything we say here.

10.1 Emission Background Subtraction and Intensity Correction

We start with the interferogram transformed into a spectrum, in whatever format is required by our reduction program. We display the desired spectral region and note its various features, such as locations of strong lines, bands, background continuum, and noise level. Figure 10.1 is the furnace spectrum in emission of ZrO in the region 10 000 cm^{-1} to 20 000 cm^{-1}. The data have been averaged to bring out the overall intensity pattern. Very strong lines do not stand out in this representation.

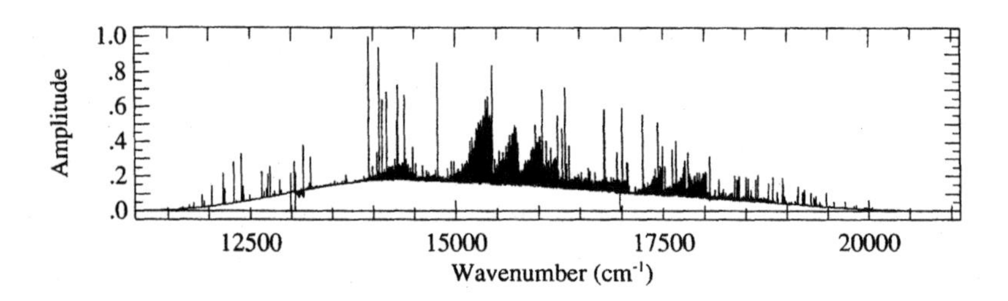

Fig. 10.1 ZrO furnace emission spectrum.

There are atomic emission lines, quite a few molecular bands degraded to the red, a few weak absorption bands of oxygen at about 13 000 cm^{-1}, and an underlying continuum. The next task is to subtract the continuum. We make a spectrum that represents the continuum by slicing up the file into sections with equal numbers of points, picking the smallest-intensity point in each section and passing a curve through these minima. This background curve is filtered or smoothed if necessary to remove any remaining local structure or artifacts and is then subtracted from the raw spectrum. In some cases, it may be necessary to perform an iteration of background subtraction before the spectrum has a level baseline. The result looks like Fig. 10.2.

Next, the instrumental transmission function is ratioed with the spectrum. To get the function, we need the calibrated spectrum of a standard lamp and the observed spectrum of this lamp made under the same experimental conditions as the unknown spectrum. The general procedure is: Call up the data file of the *observed* standard lamp spectrum; call up the *calibrated* data file of this lamp; ratio the two files to get the transmission function of the spectrometer; ratio this function with the unknown spectrum data file. We use the ZrO spectrum as the unknown data file and a ribbon filament lamp as the standard. The calibrated lamp spectrum, observed lamp spectrum, and their ratio — the instrumental transmission function — are shown in Fig. 10.3, the final corrected ZrO spectrum is shown in Fig. 10.4.

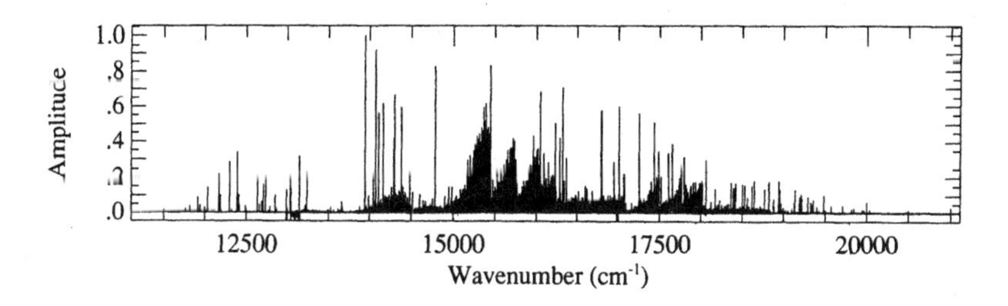

Fig. 10.2 ZrO furnace spectrum with background subtracted.

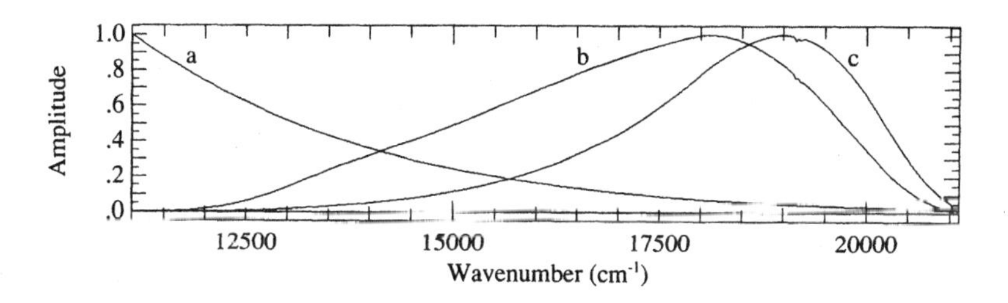

Fig. 10.3. (a) Calibrated ribbon filament lamp spectrum; (b) observed lamp spectrum; (c) instrument transmission function (ratio of b to a). All curves are normalized.

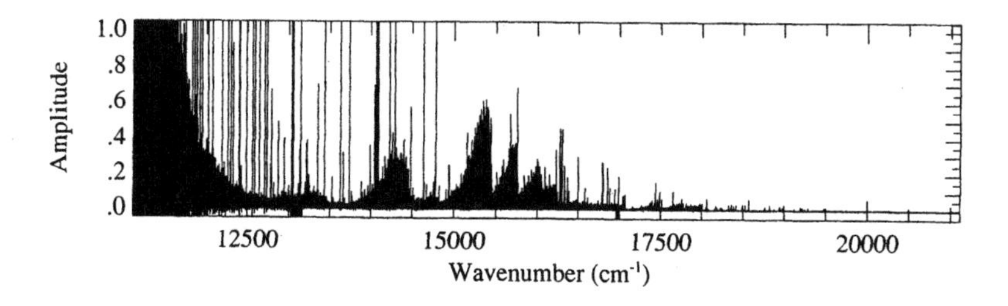

Fig. 10.4 ZrO furnace spectrum corrected for instrumental transmission.

Much care must be exercised here, because large ratios at the ends of the spectrum, where the lamp spectrum is weak, magnify the noise to such an extent that the spectrum is unusable. The S/N ratio, following transmittance correction, is *no longer independent of wavenumber.* We arbitrarily use only that part of the spectrum where the instrumental transmission is greater than 10%, in this case between 12 500 and 20 000 cm^{-1}

In some cases the calibrated lamp is not solely a smooth continuum but may have some line structure. If so, the lamp must be observed under the same spectral resolution as when it was calibrated.

The next step is to plot the spectrum at whatever amplitude and dispersion best present the features we are interested in. When there are very strong atomic lines mixed with weaker and more uniform band spectra, we can make a linear/logarithmic plot to boost weak features and reduce strong ones. In this kind of plot, the weak features are presented with a linear intensity scale, while strong features have a logarithmic scale. The two scales intersect at an arbitrarily set intensity of 10, where both the values *and* the slopes are matched. An example is the ZrO spectrum shown in Fig. 10.5.

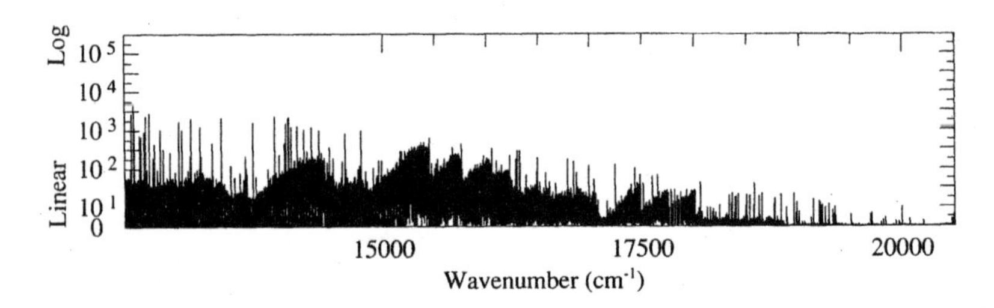

Fig. 10.5 Linear/logarithmic plot of the ZrO furnace spectrum. Features with relative amplitudes less than 10 are plotted linearly; all other features are plotted on a logarithmic scale.

Low-dispersion plots are good for seeing large-scale features such as band heads, while higher-dispersion plots accentuate band origins and individual branches, and still higher dispersion shows line shapes and blends. Examples are given in Fig. 10.6, which illustrate the peculiar nature of high resolution spectroscopy. At each level of dispersion different information is available: from inter-band structure and similarity at low disperion, to intra-band struture and the relative strengths of isotopic bands at intermediate dispersion, and the still unresolved line structure at high resolution.

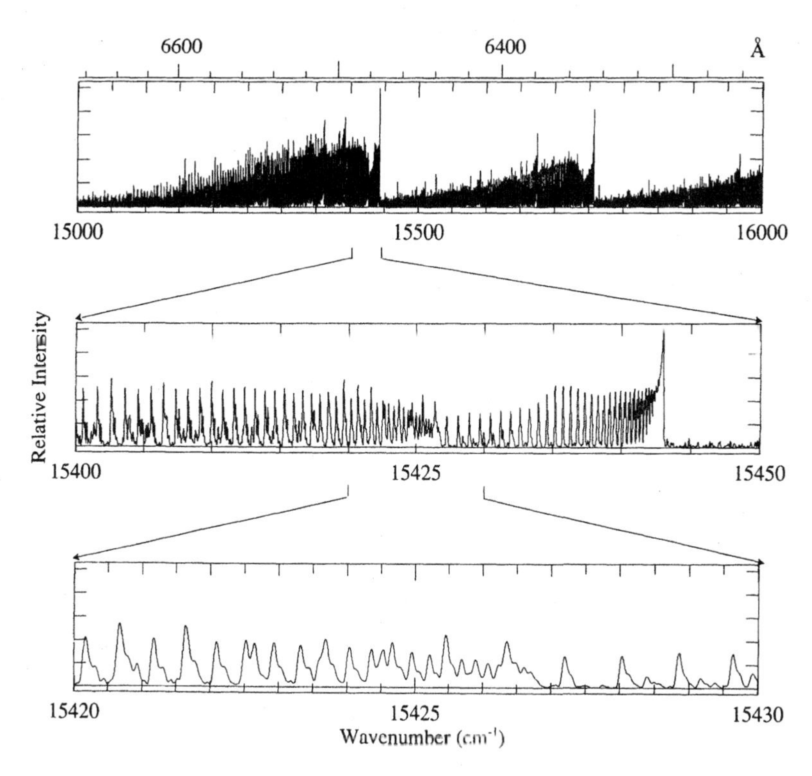

Fig. 10.6 ZrO furnace spectrum at dispersions of 118 cm^{-1}/cm, 5.4 cm^{-1}/cm, and 1.08 cm^{-1}/cm.

10.2 Absorption Background and Intensity Corrections

Let's now turn to absorption spectra, and again start by assuming we have on hand the raw spectrum. We display the desired spectral region, and again note its various features, such as locations of strong lines, bands, background continuum, and noise level. Figure 10.7 shows the absorption spectrum of CaH in a furnace. The continuum was provided by a halogen lamp.

Setting the background radiation for an absorption spectrum can be exceedingly tricky, unless a known continuum source is used. There is no problem when the absorption is weak and consists of discrete lines. But where lines are blended to create wide absorption regions, or where they are so strong that nearly 100 % of the radiation is absorbed, it is easy to make serious errors in setting the background. In the present case we do have the spectrum of the halogen lamp, as shown in Fig. 10.8. In astronomical spectra we never have the spectrum of the background radiation.

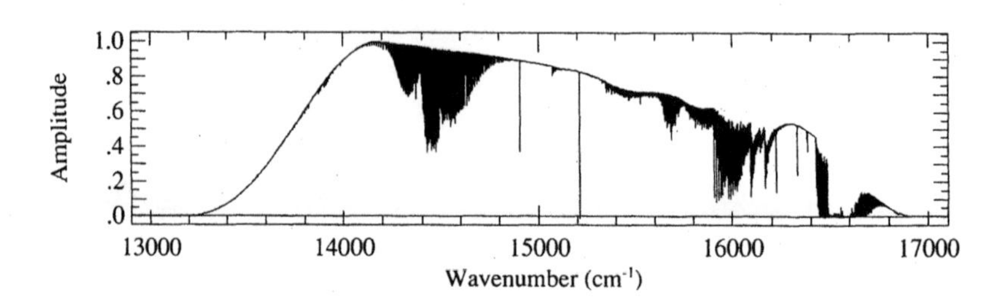

Fig. 10.7 CaH absorption spectrum.

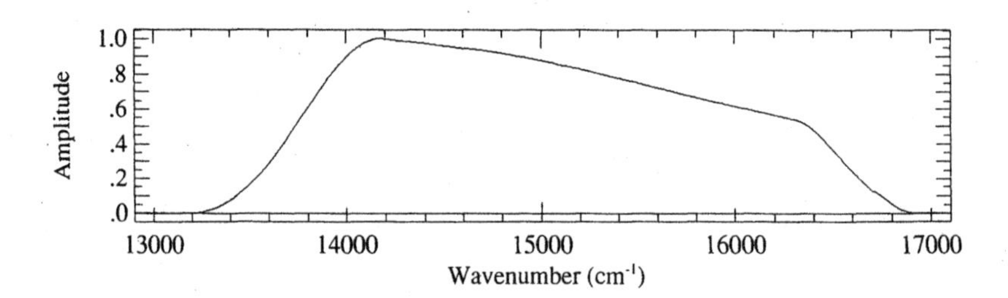

Fig. 10.8 Halogen lamp spectrum.

We ratio the absorption spectrum with the lamp spectrum, to produce a background of uniform intensity. When this background intensity is set equal to 1 we have a *transmittance spectrum* such as that in Fig. 10.9. Plots can now be made with whatever dispersion is appropriate, as in the case of emission spectra.

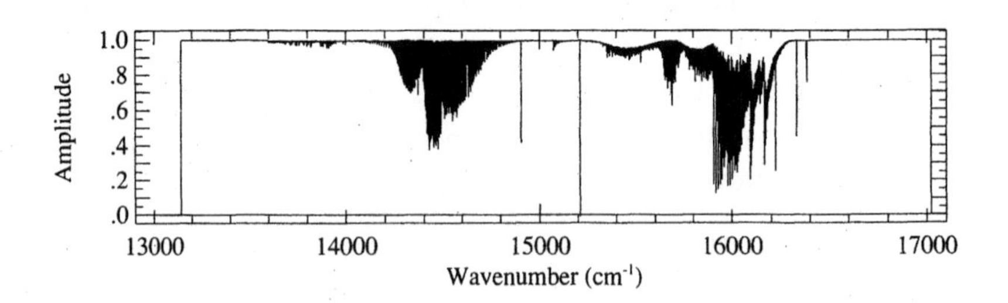

Fig. 10.9 Absorption spectrum with continuum correction, shown in transmittance.

To use the spectrum for width and area measurements, it is necessary to take the data reduction one step further. The radiation transmitted is proportional to the negative exponential of the absorbance, modified by the instrumental function. It is this exponent that is voigtian in shape. We use our code to fit the absorbance spectrum formed by taking the natural logarithm of the observed spectrum. The fitting is not completely accurate in determining the absorption coefficient, because the instrument function is convolved with the absorption spectrum rather than the absorbance spectrum. An absorbance plot is shown in Fig. 10.10. An alternative is the more complex process of calculating the absorption spectrum and convoluting this with the instrument function (ILS), and then adjusting the line parameters to fit the resulting observed spectrum. Various approaches to line parameter retrieval with fully simulated absorption lines, where the voigtian is convoluted with the instrumental line shape function, have been developed (see especially the papers of John W. Johns and Linda R. Brown) and applied to both laboratory and atmospheric measurements.

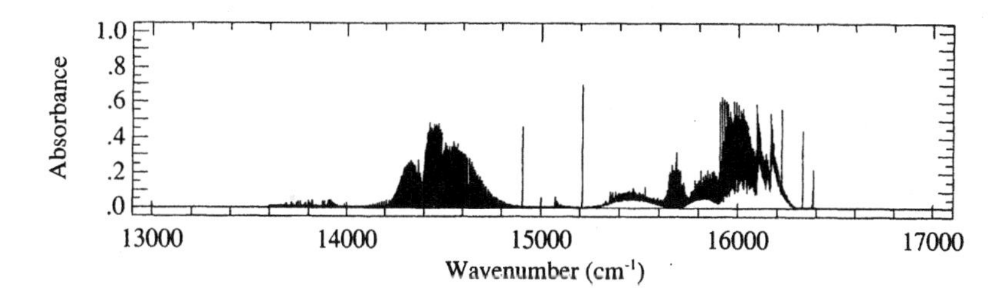

Fig. 10.10 CaH absorbance spectrum.

10.3 Line Lists and Line Fitting

With the "linelist" command and subsequent least-square fitting, we can make a list of lines and their fitting parameters directly from an emission spectrum or an absorbance spectrum. We will use an emission spectrum of CN for an example (Fig. 10.11).

First get a section of the spectrum on the screen and try picking lines with a "findlines" command. Adjust the line-finding parameters until you are satisfied that the weakest useful lines are picked while the noise is excluded.

Remember that the usual line-picking procedure evaluates the peak intensity of a line and its sharpness. A sharp weak line may be picked before a wider and stronger one. The level of picking should be set to pick most of the lines desired or a few more. Then individual lines can be added or subtracted as necessary.

When you are satisfied with the level of picking, execute the "linelist" command and generate a line list. This command assumes that all lines are nearly gaussian and assigns the same damping parameter to all of them. It is a preliminary least-square fit with a fixed value of damping. The lines are fitted with only one iteration, and then the parameters are entered into a permanent line list file.

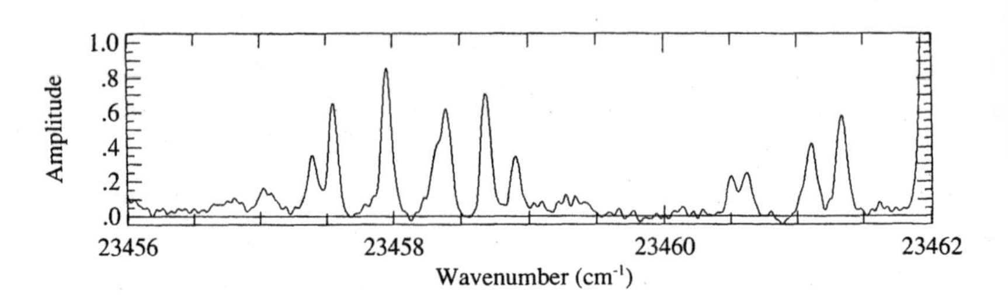

Fig. 10.11 CN spectrum to be fitted.

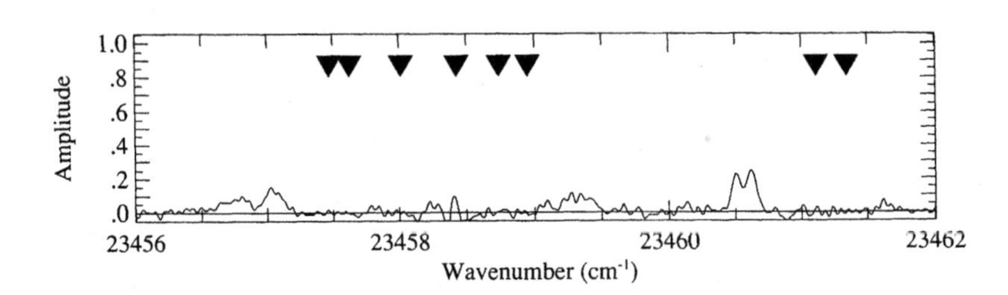

Fig. 10.12 Residuals after lines of Fig. 10.11 were picked with too high a threshold. A blended line to the left of center and a doublet at the right are not picked and are evident in the residuals.

The next step is to fit the lines to a model, in our case the voigtian profile, using the least-square method. In the fitting code, the position, intensity, width, and damping parameters are allowed to vary, but the width and damping can be set to fixed values if you have some prior knowledge of what they should be. The residuals will look similar to those in Fig. 10.12. Now patience and judgment are required. March through the spectrum of the residuals and look for lines that are poorly fitted or perhaps not fitted at all. Add or delete lines or adjust the parameters or make extra iterations when necessary.

After fitting, it is important to verify that the lines are fitted properly. Look at the residuals. With a little experience you will be able to tell when they are random and small, indications that the fit is satisfactory. You may also see lines that were not picked, and these can be added to the list and fitted individually. This process may be tedious, but it is the only way you can be guaranteed a complete and valid line list. The final residuals of a properly picked and fitted spectrum are shown in Figs. 10.13 and 10.14.

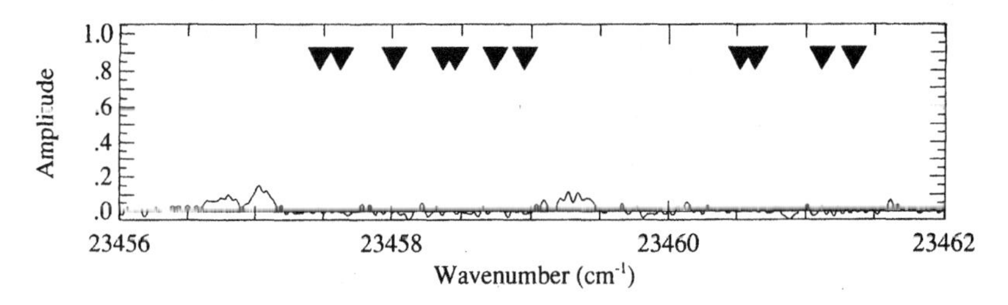

Fig. 10.13 Final residuals.

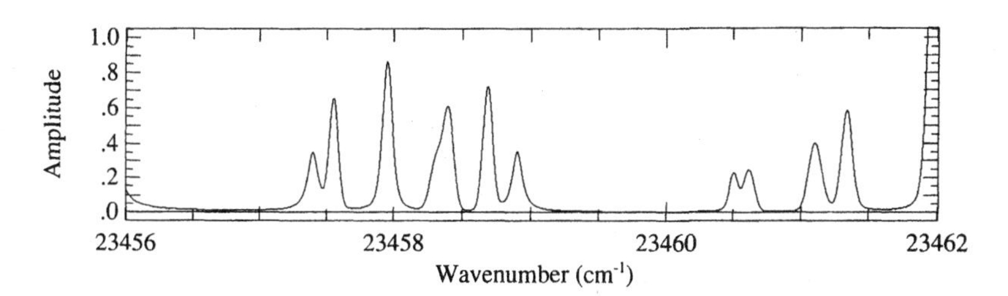

Fig. 10.14 Fitted lines only (no noise or residuals).

It is quite tempting to think that this final fitted spectrum truly represents the spectrum of the incoming radiation. Yet at each point in the data reduction we had to make a choice, and these choices influenced to some extent what the final result looks like. For example, there are end effects at the beginning and ending of the file that falsify the spectrum in those regions while affecting the center very little, as is evident in Fig. 10.14. In our illustration with the spectrum of N_2^+, should all lines have the same damping (shape)? Typically all unblended lines in the simple spectrum of a molecule do have the same shape, yet blends often result in different fitted shapes, widths, intensities, and positions. Too many iterations of blended lines can distort these parameters while making trade-offs to reduce the residuals to the smallest values without regard for the physics. The noise spectrum is essentially unknown, yet it has an influence on the line parameters. What we have done here is give you a map and instructions on how to proceed. You can start the process of data reduction with an understanding of it, but you still must build up the experience and judgment to get things right.

11

DISCUSSIONS, INTERVENTIONS, DIGRESSIONS, AND OBSCURATIONS

This chapter is an addendum, with topics that came a little late to become part of the text; it was written by Davis and Abrams for purely practical reasons, although it summarizes several of Brault's papers and unpublished works from the last decade. The material is rambling, speculative, sometimes reflecting topics either too complex or not sufficiently worked over for us to be happy with a formal treatment, full of interesting tidbits that we picked up along the way, and last of all simply a collection of things that we want to say. The Saas Fee notes concluded with, a "glance ahead and to the side": Then the notion of an imaging Fourier transform spectrometer was a tangible prediction of the future. Sixteen years later we conclude with a discussion of echelle spectrographs, which have a strange historical linkage with interferometers during the period before digital computers and also are the current necessary instrument for challenging applications previously addressed with FTS instruments.

This book began with our interpretation of Brault's Saas Fee notes. For years it and the subsequent versions of the book were referred to as the *Parum* ("too little, not enough") *Opus* with the obvious implication that the final version would be the *Magnum Opus* and with the implicit challenge that such books are usually completed after the death of the authors. In addition, the book grew out of an extended (now more than 25-year) discussion about precision spectrometry and the limitations of performance as dictated by the laws of physics. For many in the community, these discussions began on the mountain (Kitt Peak), late at night as we took data, and continued downtown at the observatory; and as often as not, the essence of the discussion would be distilled into a few lines of computer code or a revised observing strategy.

As this book approached completion, it acquired a polish that inhibited a conversational spirit (along with the loss of much chalk dust and coffee stains). In addition, as we solicited reviews and criticisms, there were some delightful opportunities to reflect on the evolution of Fourier transform spectrometry and the community of individuals who have shepherded it along. The bibliography is one such example, it grew out of Anne Thorne's insistence that it was necessary, out of the generosity of Luc Delbouille and Ginette Roland who provided the skeleton, and out of a review article on space-based FTS systems.

As the book proceeded through the review process, visitors from Giessen arrived for the holidays bearing a letter from Liège, dated a year earlier. The irony of visitors from the East arriving during Advent bearing gifts was impossible to miss. Discussions ensued, letters were written, historical references were consulted, and a group of discussions emerged that captured the spirit of our community. Some of these discussions could be included discretely in the text (a sentence here, a paragraph there), but some discussions could not be easily included without distracting the reader from the flow of the material — yet there is a certain desirability in enabling (or recording) that spirit of discussion and digression. Two subjects can be rightly viewed as advanced topics: a novel approach to sampling in interferometric systems (the "Brault algorithm") and the imaging Fourier transform spectrometer (IFTS). The logical marriage of FTS instruments and focal planes in the IFTS yields a new perspective on hyperspectral and ultraspectral imaging spectrometry. In contrast, discussions on apodization, instrumental line shapes, field-of-view effects, and data-processing techniques are as old as the field and subjects of considerable passion among practitioners.

Hence we offer this chapter of *discussions, interventions, digressions, and obscurations*. Discussions are what we do best—discuss, think, discuss some more, agree, disagree, etc. Interventions are often necessary because the participants in the previous discussion may be paradigm-impaired, whereas a third party may be able to enable (or disable) a more enlightened discussion. Fellgett's famous assertion that *apodization should be done only by experts* is a marvelous example that echoes through the decades. Digressions are the logical approach to an undesired (or inappropriate) intervention in which the party of the first part (see the preceding discussion) reclaims ownership of the discussion from the party of the third part who tried to usurp the conversation. And finally obscurations are included, because obscurity is often the court of last resort — when all else fails, leave the proof as an exercise for the student. In fact, there are two derivations that have long plagued the community, because they were anything but obvious to everyone except the inventor. As the student in charge, one of us was left with the task of documenting the derivations, but a decade later they were viewed as rather more detailed than necessary in this text. Nevertheless, they do exist in some electronic document somewhere in cyberspace.

11.1 A Novel Approach to Sampling Systems

Innovation is often the product of necessity tempered with a bit of desperation. In some cases the absence of resources leads to better approaches. In the case of sampling systems, the historical approach (as noted later) is complex and expensive. In the early 1990s Brault had the opportunity to revisit this subject, motivated by necessity of improving the performance of a balloon borne instrument and building a low-cost interferometer. The "Brault algorithm" that resulted redefines *how* one builds an interferometer, by removing the requirement for a precision drive system, a system that was the essential enabling feature of all previous FTS systems. Due to its complexity it was always a major expense and a weakness of interferometers. Among the many other insights brought to the field, the notion of replacing complex hardware with simple software borders on heresy, but totally changes how one builds interferometers. A measure of this is the fact that a "third party" has attempted to patent this concept, in contrast to the open approach taken by the inventor.

11.1.1 Introduction

We have conspicuously avoided the details of instrument fabrication — suffice it to say that the actual implementation of these ideas in glass and metal is an extended exercise in second-order effects. What is ostensibly a simple exercise in sampling the modulated radiation field on a regular basis, together with the required high spectral resolution and large free spectral range combined with radiometry that is both precise and accurate, typically require millifringe sampling and 20-bit dynamic range requirements. Each new requirement has pushed the state of the art continuously for the last four decades. In particular, two approaches have dominated FTS — step and sample, and continuous scanning with synchronous sampling. In the following discussion we will focus on rapid scanning, largely because that is the approach we have found most successful for our work.

In a continuous scanning FTS, the technical challenges reduce themselves to two basic issues: *when* to sample, and *how* to sample the interferogram. The *when* involves being able to recognize the positions of equal path difference promptly and accurately. Laser fringes have been used to establish the x-axis in interferometric spectrometry. The *how* involves the detection and measurement of the incident radiation field and includes a detailed consideration of the detectors, amplifiers, antialiasing filters, digital filters and analog-to-digital converters. The need for high dynamic range, good linearity, and low noise imposes significant restraints on the design of the sampling system.

11.1.2 Position Measurements

Fourier transform spectrometers have long been noted for stringent requirements on sampling position accuracy, and for good reason. The requirement stems from the desire to obtain a spectrum that does not contain ghosts of sufficient strength to noticeably distort the spectrum. If the system is subject to a periodic-sampling position error with an amplitude that is a fraction of the sampling interval, then the derived spectrum will show ghosts on either side of every spectral line; the strength of the ghosts is roughly that same fraction of the parent line strength. In a typical system, we may choose to keep the ghosts' strength below 10^{-3} of the parent line, which will require periodic position errors to be kept below 10^{-3} of the sampling interval. Given that the sample interval is typically of the order of a wavelength of visible light (the laser metrology fringes), this *millifringe* requirement is extremely stringent.

The situation is even worse for the precision of the drive speed. For an assumed constant sampling rate, the relative speed error must be smaller than the position errors by a ratio of $2\pi f_e/f_s$, where f_e is the frequency of the speed error and f_s is the sampling frequency. In a simple case, if the error is at 30 Hz and the sampling frequency is at 2 KHz, the relative speed error must be 10 times smaller (10^{-4}) — obviously an unachievable requirement.

For many existing systems the speed variation is actually 10 to 100 times larger than acceptable, and consequently the sampling is decoupled from the clock and triggered based on position information from the laser metrology signals (typically, laser fringe crossings). Such fringe sampling systems can achieve high sampling accuracy but at the cost of a highly restricted set of free spectral ranges that are tied to the laser frequency ($\sigma_l/2, \sigma_l/4, \sigma_l/6, \ldots$). Solutions that offer more control over the free spectral range require subdividing the laser fringes. Unfortunately, most schemes for fringe subdivision seem to carry latent potential for periodic errors, and consequently laser ghosts have become a fact of life. For example, we can double the free spectral range by sampling both rising and falling zero crossings, but any offset in the comparators used to sense these crossings will produce a periodic error at the laser fringe frequency.

The complexity of designing a precision control of the optical path difference scan to a speed uniformity of 0.1% peak to peak is best illustrated with a specific design (1-m McMath-Pierce FTS). This level of performance has been achieved through the use of a frequency-stabilized HeNe laser for OPD speed feedback. The optical path difference (OPD) speed control is achieved through the use of a Hewlett- Packard Zeeman split helium-neon laser, which produces two beams of opposite circular polarizations, physically coincident, at two optical frequencies separated by 1.6 MHz. These two beams, separated by polarizers after the beam splitter, are sent to the two arms of the interferometer; and return at frequencies shifted, in opposite signs, by the usual Doppler effect. The beams are mixed, to produce a beat frequency of 1.6 MHz plus (or minus) one Hertz per fringe per second. That signal is kept in phase with a reference frequency, generated by a very precise and stable synthesizer programmed to produce the frequency corresponding to the desired speed (expressed in fringes per second), plus the 1.6 MHz rest frequency. As such, phase comparisons can be done approximately 1.6 million times per second, or each thousandth of a laser fringe if the speed is 16,000 fringes per second. The servomechanism controlling the instantaneous carriages speeds acts to maintain as small a phase difference as possible, producing an extremely stable speed. That stability is sufficient to enable sampling the output signal at equal time intervals, corresponding to any set of desired path difference intervals. This solves the difficult problem of fringe division, necessary for a spectrometer to work in the visible and near ultraviolet spectral regions. In addition, issues resulting from time- or frequency-dependent delays in the electronic chain are also minimized. Simultaneous sampling at both equal time and equal path difference intervals minimizes the opportunity for distortion of the spectrum in the observation process. There is very little magic in implementing such a system, just a lot of details and considerable expense.

11.1.3 Sampling Systems

A key feature of the sampling system is the synchronization of the sampling system with the metrology system, the triggering that restricts the electronics package to a particular form. In contrast, during the last two decades considerable development in clocked readouts and converters has occurred, but due to the necessity of triggering the sampling system for traditional FTS instruments these converters have not been usable. In fact, much as continuous scanning greatly simplifies the design of the servo systems for the metrology system, many of the best features of a continuous sampling system are possible because the converter deals with a continuous stream of information. Inexpensive commercial analog-to-digital converters, manufactured for the audio market, are readily available (1995). The specific advantages of representative converters are that they require no antialiasing filters (which eliminates a significant noise source and simplifies the sampling scheme noticeably) and have sufficient dynamic range (20-bit — dual-channel, which eliminates the need for gain-changing amplifiers) and are fast enough to permit oversampling and subsequent decimation.

The downside of using these converters is accommodating the fixed sampling rate, specifically, extracting a sample at a given time from a series of samples at fixed times (the times corresponding to the required zero crossings or fringe subdivisions). Ironically, a minor variant of a standard digital filter (common in most FTS systems today) will easily permit reinterpolation onto the necessary sample grid.

11.1.4 A New Approach to Sampling Systems

The simplest information regarding *when* to sample is the times associated with the laser fringe crossings (typically either positive- or negative-going crossings are used). Fringe timing can be simply measured with an interval timer with a clock rate that is sufficiently faster than the fringe rate to provide the necessary time resolution (running the clock 1000 times faster than the fringe rate will yield millifringe resolution). Hence precise fringe times $t(x_n)$ and equally spaced time interval interferogram samples $I(t_j)$ are readily available; the missing element is a method of deriving $I(x_i(t))$, where the sample interval x_i is typically a subdivision of the fringe interval x_n.

From the sampling theorem, if the fringe time $t(x)$ is a band-limited function of x and the bandlimit is less than one-half of the laser fringe rate, then knowing the times of fringe crossings is enough to recover $t(x)$ for any x. Recognizing that most instruments will be subjected to vibrations that induce speed variations in the 0–500-Hz range, a typical laser metrology system running at more than 1 KHz will be sufficient to ensure Nyquist sampling (and most laser metrology systems run at significantly higher rates). Consequently fringe subdivision reduces to a simple interpolation problem; in practice, cubic polynomial interpolation has been sufficient. Similarly, assuming the interferometric data are similarly Nyquist sampled, then the derivation of the desired interferogram values at the desired sample intervals x_i is also a matter of interpolation, which can be accomplished with a convolution of the sampled data with an appropriate continuous sinc function through the use of a digital filter (which is used to define the desired spectral alias). The added complexity of exact interpolation of the interferogram data is necessary because the interferometric signal utilizes the full bandwidth of the signal chain.

The *magic* implicit in this new approach is *how* it accommodates imperfections in the interferometer sampling system and the speed servo, in particular. As noted already, the historical approach has involved simple brute force: Build massive instruments, seismically isolate them, and utilize layers of precision servos to compensate for residual imperfections. Successful precision FTS designs skillfully handle second-order errors, i.e., those whose amplitudes depend on the product of two simultaneous errors (with the implicit acknowledgement that there are numerous *unsuccessful* and hence imprecise designs).

In the case of nonuniform-speed errors, the necessary information is provided by the fringe-time measurements: Since the fringe times are inversely proportional to the speed, they provide a direct measurement of the instantaneous speed. Reinterpolation of the interferometric signal to compensate for the speed variation is simply accomplished by adapting the digital filter kernel to have a width that is inversely proportional to the instantaneous speed. Any other stable chromatic effect (i.e., one that is a function of wavenumber σ only — such as chromatic phase error) can be corrected at the same time by a real-time operation similar to a Forman phase correction. In addition, there are components in the signal chain with frequency-dependent phase and amplitude responses that are sensitive to speed variations. The solution for compensation of speed variation is to utilize a two-dimensional array of interpolating digital filters, where the second dimension is speed. Each kernel is a convolution of two components: one that scales with speed (for functions with a wavenumber dependence) and one that is independent of speed (for functions with a frequency dependence).

11.1.5 Practical Realization, Results, and Implications

The implementation of the design resulting from these concepts is remarkably simple and was demonstrated in a laboratory instrument built by Brault and Jakoubek for NOAA at the Fritz Peak Observatory and enabled the design of the interferometer without a precision drive mechanism. A simple open-loop scanning was used based on a commercial microstepper motor and indexer. Test data taken with a laser as the source demonstrated a severely broadened line profile (the FWHM was 7% of the total spectrum) due to speed-induced ghosts when the interferometer was operated without the speed-corrected sampling. With the speed-corrected sampling system, the resulting line profile was consistent with the expected linewidth of the laser. On a relative basis the performance of the interferometer with speed correction is 1000-fold better than without speed correction.

The practical implications of this result is to redefine *how* the sampling systems of all future interferometers will be designed and evaluated. In the years immediately following the publication of these ideas and designs, numerous groups implemented some version of the *Brault algorithm.* For many of us, it has liberated our instruments from a major design constraint and challenge, both technically and financially, and ironically has made the instruments simpler and more robust at the same time.

11.2 The Imaging Fourier Transform Spectrometer (IFTS)

In recent years the combination of visible and infrared focal plane arrays and Fourier transform spectrometers has opened a new perspective on imaging spectrometry. The IFTS exploits the stigmatic imaging capability of the FTS with Nyquist-sampled imaging to enable true hyperspectral imaging. The hypercube has two spatial dimensions and one spectral dimension, or 3-dimensional imaging in the astronomical case, with the spectral axis of a spatial-spectral hypercube mapping directly into distance (red shift)). While the imaging capability is easily understood for the classical two-port FTS, the four-port FTS offers the added advantage that every photon is transferred to a focal plane. The IFTS provides a different and efficient means of conducting *unbiased* spectroscopic surveys, i.e., without object or spectral band/color preselection and without the restrictions imposed by slit geometry and placement. In addition, the IFTS also allows spectroscopy over a wide bandpass, affords flexibility in choice of resolution, is easy to calibrate, and is ideal for wide-field spectroscopic surveys.

An IFTS is axially symmetric, and the optical path difference (OPD) is the same for all the points of the source with the same angle of incidence from the axis of the interferometer. Hence, the field of view (FOV) is circular. On the object side, an entrance collimator illuminates the interferometer with parallel light. The interfering beams are collected by the output camera, creating a stigmatic relation between the object and image planes. By placing a detector array in the output focal plane, the entrance field is imaged on the array and each pixel works as a single detector matched to a point on the sky.

Retrieving spectral information involves recording the interferogram generated by the source imaged onto the focal plane array (FPA). The OPD is scanned in discrete steps, since FPAs are integrating detectors. Scanning in this way generates a data cube of two-dimensional interferograms. The signal from the same pixel in each frame forms an independent interferogram. These interferograms are Fourier transformed individually, yielding a spectral data cube constituted with the same spatial elements.

Beyond the efficiency and flexibility of IFTS instruments, there is an additional compelling reason for using a multiplex spectrometer: In the dual port design, virtually every photon collected is directed toward the focal plane for detection. Other solutions are inefficient, inflexible, and wasteful of mass, power, and volume. Cameras equipped with filters admit only a restricted bandpass at low spectral resolution. To compete with the spectral multiplex advantage of an IFTS, a camera system needs multiple dichroics and FPAs. The additional mass and thermal loads are a severe penalty. Classical dispersive spectrographs have slit losses, grating inefficiencies due to light lost in unwanted orders, and limited free spectral range (the same is true for a Fabry–Perot). An IFTS acquires full bandpass imaging simultaneously with higher-spectral-resolution data. Therefore, a high S/N broadband image always accompanies full spectral sampling of the FOV, with no penalty in integration time. An IFTS is a true imaging spectrograph and measures a spectrum for every pixel in the FOV. It is not necessary to choose which regions in the image are most deserving of spectroscopic analysis. Overheads are eliminated, because no additional observing time is needed for imaging prior to object selection, and there is no delay in positioning slit masks, fibers, or image-slicing micromirrors. An unstated, but implicit, assumption is that the potential for serendipity will enhance the scientific legacy obtained with IFTS instruments.

Two features of an IFTS instrument should be noted: (1) An IFTS is spectrally multiplexed; therefore all spectral channels are obtained simultaneously within the stated integration time. (2) The free spectral range of an IFTS is limited only by the bandpass filter and the detector response. Consequently, the usual definition of resolution, $R = \lambda/\delta\lambda$, is of limited use. It is conventional to scan the OPD of an IFTS in equal steps so that the resolution is constant in wavenumber. Thus, R is used to denote the number of spectral channels. For example, in the near-infrared (NIR) with a 1- to 5-μm bandpass, $R = 5$ means that $\delta k = (k_{\text{max}} - k_{\text{min}})/R = 1600\ \text{cm}^{-1}$, and a scan yields five bands centered 1.1, 1.3, 1.7, 2.3, and 3.6 μm.

The throughput of an IFTS is limited only by the efficiency of the beamsplitter. In a dual-input, dual-output port design, no light is wasted and the throughput approaches 100%. On blaze, a good grating is 80% efficient, but averaged over the free spectral range this drops to about 65%. An IFTS has no loss of light or spatial information because there is no slit. An IFTS is well adapted to doing multiobject spectroscopy in crowded or confusion-limited fields. Admittedly an IFTS is less efficient than a traditional slit spectrograph for single-object spectroscopy. When we ignore slit losses, blaze function and small free spectral range of gratings, a slit spectrograph is R times faster than an IFTS of resolution R in a background-limited operation.

However, *spatial* multiplexing renders the performance of an IFTS equal to an ideal multislit spectrograph, and the other advantages of an IFTS can be overwhelming. The spectral resolution can be varied arbitrarily from the coarsest case of a small number of bands up to a spectral resolution limit determined only by the maximum OPD characteristic of the instrument. An instrument with a maximum OPD of 1 cm can be operated over a range of resolutions from full band up to R = 10,000. The spectral resolution limit of a Michelson interferometer used as an IFTS is

$$R = 8(d/\phi D)^2, \tag{11.1}$$

where ϕ is the FOV, d is the diameter of the beamsplitter, and D is the telescope diameter. Classically, ϕ refers to the entire field. This is not the case for an IFTS, where ϕ is the FOV of an individual pixel. Although convenient that a single fringe should fill the FPA, just as with imaging Fabry–Perots, there is no reason why each pixel should record the same apparent wavenumber. For an IFTS the size of the optics is determined not by spectral resolution, but by the requirement that there be no vignetting. Therefore the optics for an IFTS are similar to those of a simple reimaging camera and smaller and slower than those of an equivalent dispersive spectrograph.

An IFTS is tolerant of detector noise because it always operates under photon-limited conditions due to the broad spectral bandpass transmitted to the FPA. Similarly, orders-of-magnitude higher thermal emission from the instrument, or thermal radiation leaks from outside the instrument bay, can be tolerated compared to the case for dispersive spectrometers or fixed-filter cameras. An IFTS is also tolerant of cosmic ray hits, because the "energy" of a single upset pixel in one OPD frame divides among all bins in the spectral transform of the interferogram for that pixel.

A dual-port design delivers the complementary symmetric and antisymmetric interferograms. The final interferogram is constructed from the difference (which is therefore also immune to common-mode electrical noise), while the normalized ratio reveals systematic variation due to detector drifts.

The wavelength scale and the instrumental line shape (a sinc function if there is no apodization) are precisely determined and are independent of wavelength. Absolute wavelength calibration is done by counting fringes of an optical single-mode laser. Compared to a dispersive system, the broadband operation of an IFTS means that there are R times more photons for flat-fielding and determining signal-dependent gain (linearity). Hence, high signal-to-noise calibration images can be acquired faster or with lower-power internal sources.

11.3 Characterization and Determination of Instrumental Line Shape Functions

The characterization, diagnosis, correction, and improvement of instrumental line shape functions has been the never-ending focus of our lives for most of our careers (which is how one stays employed as a high-resolution spectroscopist!). Chapter 5 may be considered the *tip of an iceberg*, and certain additional warnings should be conveyed for the unfortunate reader who has to try and understand why a given instrument does not meet the expected performance metrics. At the center of this discussion is the issue of field- of-view effects and the indicators of when ILS distortions are inconsistent with correctable effects.

The impact of oblique rays most typically appears when the illumination of the field of view does not have circular symmetry. When this happens, a single-sided "shoulder" appears in the instrument profile. A related problem affects some instruments: If the off-axis optics are poorly made or poorly aligned, they can introduce aberrations that degrade the field-stop image. The result is a distortion of the nominal rectangular part of the instrumental profile that produces another "shoulder" on the ILS, always on the same side. Perhaps the best advice is to understand the ILS in detail and to assess the alignment and illumination regularly. In such a fashion these second-order effects can probably be avoided by proper design or adjustment of the instrument.

When the field of view is expanded, an interesting possibility arises: Is it possible to project the output image onto multiple detectors and increase the throughput? There are two approaches, an integrating optic that maps light from a concentric ring of the image onto a linear or rectangular detector, or simply an imaging array detector, summing the individual signals from pixels in concentric circles to obtain a series of interferograms. Following transformation and "de-stretching" to account for the change in optical path difference as a function of angle, the spectra could then be coadded to maximize the spatial multiplexing possible in a stigmatic imaging spectrometer. Such a concept (first proposed to us by Delbouille and Roland) would take the Jacquinot advantage to a new level, with pixels situated in concentric circles to get a series of interferograms.

11.4 Apodization

Among spectrometers, the FTS is unique in offering the opportunity to operate near minimally sampled. (One may reconsider Fig. 2.2, which indicates that the necessary sampling for 1% error is two to three samples for an FTS and five to ten for a grating spectrometer). Not surprisingly, many users eventually run their spectrometers minimally sampled and then find that they need yet more resolution to resolve the features in the spectrum. At this point the observed spectrum is heavily distorted by the instrumental line shape (ILS) profile. There are four techniques to address the distortion of the spectrum: (a) buy a bigger spectrometer, (b) numerically filter the spectrum (apodization), (c) fit the instrumentally distorted spectrum in a fashion suitable to extract accurate estimates of the spectrum, and (d) remove the instrumental distortion in a statistically deterministic process. Obviously the first solution is both trivial and irrelevant to address the issue of extracting the best results from a given data set. The remaining three solutions are variations on the theme of using computational methods to address the limitations of instruments.

11.4.1 Introduction

The challenge of achieving a suitable compromise between sidelobe attenuation and line shape distortion has been a longstanding debate within the FTS community. The specific application of numerical filtering to the alteration of the ILS function has been termed *apodization*, stemming from the notion of removing the oscillations (or feet) from the ILS function (The word *apodization* was introduced by Jacquinot in 1950 to describe "the process of modifying the diffraction figure in such a way to attenuate its feet."). As a result of two key words — *suitable* and *compromise* in the previous sentence, the particular or recommended implementation remains largely a matter of opinion. Unfortunately, much of the work within the FTS community was performed with little contact or exchange of ideas with the signal processing or electrical engineering communities. For example, the problem of minimizing the ILS sidelobes is mathematically identical to the process of creating a highly directional radar antenna. Hence, two (or more) solutions were invented, named (of course), and propagated by purveyors of solutions in a relative vacuum. Occasionally, individuals would compare the relative performance of various apodization functions, establish metrics, and recommend particular solutions for particular applications.

Two historical notes are worth mentioning. In the beginning it appears that the apparent analytic simplicity of the triangle function made it "look easier to compute". One may find numerous examples of computations that produced erroneous results just because the researcher liked the simplicity of the algorithm. Second, we have strongly urged that apodization be avoided at all costs. In this we must confess to being slightly schizophrenic — the key is achieving an intimate understanding of *how* the data are being distorted in the processing and assessing the impact of the distortion on the interpretation of the results. As a general rule, apodization, and certainly routine or naive apodization, must be discouraged. The literature is rampant with data that have been mangled, destroyed, or misinterpreted due to apodization. The worst offenders utilize apodization blindly as implemented in canned software and "look" for the best spectrum visually. The best cases throw away 50–60% of the information contained in the interferogram without considering the incremental gain or loss of information in the process.

Conversely, apodization is a primary method of addressing the finite signal-to-noise ratio of both the data and the throughput of our numerical algorithms. An excellent example is numerical differentiation, which typically degrades the signal-to-noise ratio by a factor of 10 with each derivative. Careful selection of numerical filters dramatically improves the performance of the differentiator without significantly distorting the results — the key words are *improves* and *significantly*; both require quantification to define the relative performance.

P. Fellgett is notorious for a sharply worded intervention on apodization: "the orthogonal properties of the sinc function are easily destroyed by apodization; that is why I believe that apodization should be done only by experts." (Orsay, 1967). He was annoyed to see colleagues always apodizing their data, and, in doing so, passing from the best possible instrumental profile (in the stated conditions) to something less favorable (mimicking, for example, grating instruments but very often doing worse.) Thirty-five years later his concerns remain valid and infrequently observed.

Over the last 40 years, a considerable amount of work has been performed on apodization as specifically applied to FTS: Filler in the 1960s, Norton and Beer and Harris in the 1970s, and Brault in the early 1990s. In the beginning each group used its own apodization functions, attenuating the sidelobes of the instrumental profile at the cost of its FWHM according to various compromises. Filler proposed plotting the performance of each described function on a diagram connecting the ratio of the instrumental FWHM before and after apodization with the amplitude of the first sidelobe, normalized on the main peak amplitude. Filler suggested that there is a boundary under which no better solution will be found. Norton and Beer proposed three apodization functions falling practically on that boundary, and commented on the general form of optimal solutions. In the same year, Harris presented a comprehensive analysis of apodization functions, in the context of windowing for signal processing. Brault, in an unpublished work, unified the previous literature in a comprehensive performance analysis.

In spite of the literature, many applications continued (and continue) to use the primitive triangle apodization function, which is simple but produces terrible results. An intermediate group of applications use the Norton–Beer functions (either by choice or because they are included in the package purchased with their instrument) in a relatively routine manner. A third group of applications rigorously avoids the use of apodization, although the impact on the quality of the results is rarely evaluated. Finally, some applications have conducted detailed performance analyses and tailored an apodization scheme to the particular application and described the error analysis in detail.

11.4.2 Selected Apodization Functions/Windows

11.4.2.1 Rectangle/Dirichlet Window

The rectangle function is unity over the observation interval and zero thereafter and is the literal representation of the truncation function for the finite path difference of an FTS instrument:

$$W(n) = 1.0. \tag{11.2}$$

The Fourier transform of this function is the sinc function (or Dirichlet kernel) and the canonical instrumental line shape function for an FTS instrument. The magnitude, oscillatory sign, and rate of decay (which is very slow) of the sidelobes of this function have led many groups to use some form of apodization to attenuate the sinc function into something more acceptable. Unfortunately, in many cases the input interferograms have not been retained, and hence reprocessing the data with other apodization functions is impossible. It is worth noting that the locations of the zeros of the sinc function define the intervals between statistically independent samples, and the use of any other apodization function will alter the independence of the samples and shift the locations of the zeros.

11.4.2.2 Triangle Window

The triangle function is the auto-correlation of the rectangle function and is the classical tapering function used in early FFT work to reduce the computational time (with the implicit caveat that the distortion of the signal was acceptable):

$$W(n) = \begin{cases} \dfrac{n}{N/2}, & n = 0, 1, \ldots, \dfrac{N}{2} \\ W(N-n), & n = \dfrac{N}{2}, \ldots, N-1. \end{cases} \tag{11.3}$$

The transform of the triangle function is a sinc^2 function, which is all positive and, despite its frequent use, has little to justify its use except precedent.

11.4.2.3 Triangle Squared Window (Riesz)

The triangle squared, or Riesz, window

$$W(n) = 1.0 - \left|\frac{n}{N/2}\right|^2 \tag{11.4}$$

is the simplest continous polynomial window. It has a discontinous first derivative at its boundaries, and hence its transform falls off as the inverse square, leading to a large distortion of the FWHM for the sidelobe suppression achieved.

11.4.2.4 Cosine Squared (Hann/Hanning) Window

The cosine squared, or Hann, window is a continuous function with a continuous first derivative, and hence its transform falls off as the inverse cube; while an improvement over the triangle and triangle squared, it is still far from optimum.

$$W(n) = \cos^2\left[\frac{n}{N}\pi\right] \tag{11.5}$$

11.4.2.5 Hamming (Happ–Genzel) Window

The Hamming window is a special case of the Hanning window, with carefully selected coefficients to ensure proper cancellation of the sidelobes.

$$W(n) = 0.54 - 0.46\cos\left[\frac{2\pi}{N}n\right] \tag{11.6}$$

A closely related function achieves minimum sidelobe suppression, but the historical definition of the Hamming window is associated with these particular coefficients.

11.4.2.6 Blackman/Blackman–Harris Windows

The Blackman and Blackman –Harris windows are generalized sums of cosine functions with carefully choosen coefficients:

$$W(n) = \sum_{m=0}^{N/2} (-1)^m \alpha_m \cos\left[\frac{2\pi}{N} mn\right], \tag{11.7}$$

subject to the constraint that

$$\sum_{m=0}^{N/2} \alpha_m = 1.0. \tag{11.8}$$

The Hanning (cosine squared) and Hamming functions are windows of this form, with α_0 and α_1 nonzero. The exact Blackman window is a three-term window with coefficients

$$\alpha_0 = \frac{7938}{18608} \approx 0.42659071 \sim 0.42 \tag{11.9a}$$

$$\alpha_1 = \frac{9240}{18608} \approx 0.49656062 \sim 0.50 \tag{11.9b}$$

$$\alpha_2 = \frac{1430}{18608} \approx 0.07684867 \sim 0.08, \tag{11.9c}$$

and the window obtained with the two-place approximations is the "approximate" Blackman function (although in common practice, the "approximate" is dropped). The Blackman–Harris functions are three- and four-term windows of the form

$$W(n) = \alpha_0 - \alpha_1 \cos\left[\frac{2\pi}{N} n\right] + \alpha_2 \cos\left[\frac{2\pi}{N} 2n\right] + \alpha_3 \cos\left[\frac{2\pi}{N} 3n\right]. \tag{11.10}$$

11.4.2.7 Norton–Beer Functions

In 1976, Norton and Beer evolved a family of apodization functions that have become rather popular within the FTS community. The functions are specialized sums of polynomial functions:

$$W(n) = \sum_{m=0}^{N} \alpha_m (1 - n^2)^m, \tag{11.11}$$

subject to the constraint that

$$\sum_{m=0}^{N} \alpha_m = 1.0. \tag{11.12}$$

11.4.2.8 Dolph–Chebyshev Function

In antenna design, the analogous problem is the illumination of an aperture to achieve a narrow main-lobe beam pattern while restricting the sidelobe response. The closed-form solution to the minimum main-lobe width for a given sidelobe level is the Dolph–Chebyshev window. The window is defined in terms of uniformly spaced samples of the windows Fourier transform $W(k)$:

$$W(k) = (-1)^k \frac{\cos\left\{N \cos^{-1}\left[\beta \cos\left(\pi \frac{k}{N}\right)\right]\right\}}{\cosh\left[N \cosh^{-1}(\beta)\right]}, \qquad 0 \le k \le N \qquad (11.13)$$

where

$$\beta = \cosh\left[\frac{1}{N} \cosh^{-1}(10^{\alpha})\right] \qquad (11.14)$$

and

$$\cos^{-1}(x) = \begin{cases} \frac{\pi}{2} - \tan^{-1}[x/\sqrt{1.0 - x^2}], & |x| \le 1.0 \\ \ln\left[x + \sqrt{x^2 - 1.0}\right], & |x| \ge 1.0 \end{cases} \qquad (11.15)$$

where α is the log of the ratio of the main-lobe level to the sidelobe level and the term $(-1)^k$ alternates the sign of the samples to reflect the shift in the origin in the time domain. The window samples are obtained by a Fourier transform on the samples $W(k)$ and scaled for unit peak amplitude.

11.4.3 Evaluation Criteria and Relative Performance

The challenge in apodization of FTS data is simple, yet surprisingly difficult: Eliminate the oscillatory sidelobes in the ILS function with minimum distortion of the ILS FWHM in a mathematical form that lends itself to rapid calculation. However, such a multivariable challenge leaves plenty of opportunity for divergent solutions. In practice, simple algebraic formulas have been developed that are representative solutions but that do not allow the user to define the level of distortion. Furthermore, the absence of parametric families makes the selection process rather binary, as indicated in Fig. 11.1. For example, the triangle, Hamming, and Blackman functions are discrete points in the performance trade space: 5% sidelobe leakage at 50% FWHM distortion vs. 0.08% sidelobe leakage with 55% FWHM distortion vs. 0.015% sidelobe leakage with 90% FWHM distortion. If you desire 1% leakage or 25% FWHM distortion, then you must design your own function.

The Filler diagram permits the comparision of two or more functions, and there appears to be a performance boundary that no family of functions crosses. The form of that boundary would represent one approach to optimal apodization. Strangely, despite challenges in the mid-1960s and 1970s, an analytic expression for the family of functions that defined the boundary was never defined. The empirical boundary and the previously mentioned notion that there is no familiy of functions that crosses the boundary ignored a key family of solutions; the Dolph–Chebyshev polynomials are the closed-form solution to the minimum main-lobe width for a given sidelobe level. Unfortunately, the penalty is that these minimum sidelobes go on forever at the same amplitude, which is a bit ugly. As a result, the Dolph–Chebyshev functions are not suitable for the processing of FTS data.

11.4.4 Families of Functions

In addition to the classical windows, Fig. 11.1 raises an interesting question regarding families of functions: Can an optimization strategy be devised that allows the selection of either a suppression factor or a distortion ratio for a given family of functions? In a simplistic fashion, the Norton–Beer functions are polynomial constructions, and the Filler, Blackman, and Blackman–Harris functions are sums of cosine functions, but they are nonetheless computed for discrete cases — not the general problem of optimal filtering for a given sidelobe suppression.

11.4.5 Method of Optimization

Apart from iterative exploration, the process of optimization is straightforward. With three parameters to be determined with two constraints from the normalization and the desired FWHM, there is only one parameter left to vary in order to minimize the sidelobes. When such a process is applied to the Norton–Beer functions, the performance of W_{NB1} improves by 7%, and W_{NB2} improves by 4% and even drops beneath the Norton–Beer empirical boundary. Typically, the apodizing terms are normalized to 1.0, and the ILS terms are whatever the FFT gives, but this is not critical.

11.4.6 Some Useful Functions for Optimal Apodization

Among a myriad of possible families, not surprisingly the three- and four-term Blackman–Harris functions (labeled 1-3-5, and 1-3-5-7, for simplicity) provided the most desirable results (for historical completeness, these functions are a modification of Filler's E_α function and termed are P_α by Norton and Beer). As shown in Fig. 11.1, the three-term optimal apodization functions are the best single family we have found, covering the interval between $W/W_o = 1.0$ to 1.88, often below the N–B boundary and better than the Norton–Beer functions W_{NB1} and W_{NB3}.

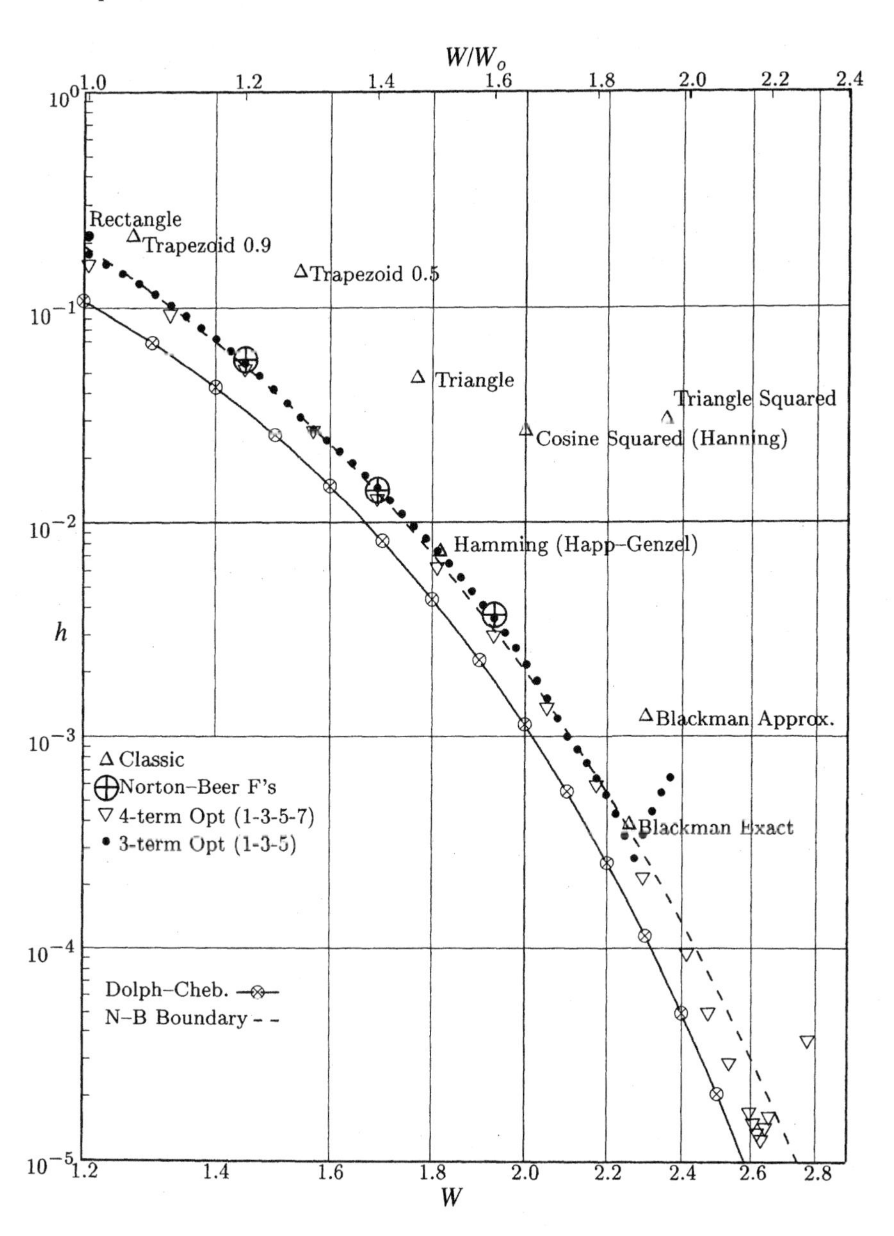

Fig. 11.1 Filler diagram for a variety of classical windows (rectangle, trapezoid, triangle, triangle squared, and Hanning), constructed windows (Hamming, Blackman, Norton and Beer, and Dolph–Chebyshev), and two families of optimum windows (1-3-5 and 1-3-5-7). The vertical axis (h) describes the sidelobe suppression, and the horizontal axis describes the width of the instrumental line shape function after apodization (W). For simplicity of use, the ratio of widths (W/W_o) on the top scale provides a direct estimate of the relative distortion of the FWHM by the apodization process.

As noted, earlier, the Hanning and Hamming functions are two-term versions of this family, and in fact there are two of the optimum 1-3-5 functions that have C5 = 0, and hence are actually 1-3 functions as well. The Hamming function is remarkably close to one of these optimal functions and has remarkably good properties for a two-term representation. Norton and Beer deride the Hamming function as "far from [the] 'best' function," a conclusion that is wholly inconsistent with its location on the Filler diagram.

11.4.7 Discussion

Among the classical functions, the trapezoids, triangle, and $\cos^2$ (Hanning) are unacceptable, in that the FWHM distortion is large for the relative sidelobe suppression. The triangle squared is by far the worst of the classical windows. The Hamming (Happ–Genzel or Blackman two-term) function is quite good, as is the Blackman three-term (exact), although both can be slightly improved with optimization.

For a wide-range optimum function, the optimized 1-3-5 family looks hard to beat; it stays remarkably close to the boundary all the way up to $W/W_o = 1.9$, and is certainly better than Norton 1-3-5 past 1.6. The optimized 1-3-5-7 (four-term) family is consistently below the empirical boundary and extends to $W/W_o \sim 2.15$. Gaussian functions are relatively disappointing, though they are not too bad for functions with no adjustable parameters. It is also the only game in town above $W/W_o = 2.0$.

11.4.8 To Apodize or Not to Apodize — Case Studies

The overwhelming question is "when, if ever, should one apodize a spectrum?" The immediate reactions of "never" or "use sufficient resolution to 'properly resolve the spectrum'" dodge the issue of making the best of a given data set and understanding the objectives in representing and processing the data. Practically, the answer depends on the goal of processing the data: When creating an atlas for visual inspection, some apodization can moderate the ringing in the spectrum and permit more comfortable inspection; conversely, algorithms in general desire the full bandwidth of information but are highly sensitive to noise, and hence providing excess bandwidth is counterproductive.

In the historical archives there are numerous examples of data sets where the interferograms were not retained and only apodized spectra retained. The use of the triangle apodization function in many cases literally means that 50% of the spectral resolution (and hence information) has been lost in the data processing. Conversely, there is an interesting challenge presented by broadband (5- to 15-micrometer) high-resolution atmospheric data: At the long-wavelength end of the spectrum there is insufficient resolution to resolve the lines, and at the short-wavelength end of the spectrum there is too much resolution and hence the signal-to-noise ratio is lower than it could be. In such a case judicious use of apodization will improve the spectroscopic measurements obtained from the spectrum, but the impacts (both positive and negative) need to be quantified and documented, with considerable care. However, the spectra themselves should not be apodized; rather, apodization may be included in the fitting process, where it can improve the signal-to-noise ratio without distorting the measured parameters. In practical applications, the simultaneous apodization of observed and calculated spectra is a method of controlling the information bandwidth that the fitting algorithm has access to and hence of biasing (as necessary) the fitting process toward a particular class of solutions. This is especially critical in cases where the signal-to-noise ratio is low and the likelihood of fitting noise spectra as real features is sufficiently high that one cannot ignore it. Notice that apodization in a fitting program is signficantly different than traditional apodization: In many cases there are undiagnosed features in the measured data (or inadequacies in the model spectra) that we cannot remove but that we know from experience are not relevant to the measurement process.

The key to proper utilization of apodization is testing: demonstrating, evaluating, and quantifying the impact of numerical apodization on the process and documenting the decision. Conversely, blind use of canned algorithms will likely produce spectra that look like they were processed with canned algorithms. Because the field of FTS has evolved and grown, there are now commercial instruments with capabilities similar to (or superior in particular parameters to) research instruments; however, there are no examples of commercial processing-and-analysis packages that offer the user ready insight into the processing of the data and the opportunity to understand the form and quality of the data. Ultimately, each researcher must demonstrate the sufficiency of their total processing (observation and measurement) to the researcher's satisfaction.

11.5 The Quest for the Perfect Instrument

The 1-m FTS at the McMath–Pierce Telescope of the National Solar Observatory has been a critical component of our careers. BABAR (the white elephant in the children's stories by Jean de Brunhoff, to those with a personal relationship with the instrument) and its particular attributes of resolution, free spectral range, throughput, stability, and reproducibility enabled us to focus on the spectroscopy of atoms and molecules rather than on explaining artifacts and imperfections due to the features of a particular instrument. In a word, BABAR was the perfect instrument for us. The combination of high resolution and a very small field of view meant that most spectra where close to ideal; and the signal-to-noise performance was essentially consistent with theory (a real measure of an instrument design). Over 25 years after first light, many spectroscopists still make the pilgrimage to the mountain to take spectra.

For perspective, our laboratory research at Berkeley was in the doldrums in 1975, when we first went to Kitt Peak. Some of our previous spectral observations obtained using grating instruments were the standard in the astronomical community and still are, but the data produced at Kitt Peak were perhaps an order of magnitude better in some cases. Even today, one of the most demanding younger spectroscopists still continues to use BABAR, even though he owns a commercial instrument.

In recent years, echelle spectrographs have come into wider use because of their speed and two-dimensional spectral displays that make possible the recording of many data points simultaneously in two dimensions rather than one. For medium-resolution spectra they are practical and satisfactory, because they are fast and convenient. The HIRES instrument at the Keck telescope represents the best of the echelle instruments and has produced data far superior to any in existence for astronomical spectra.

At higher resolutions there are some data analysis problems that are difficult and practically impossible to solve. Each order covers a relatively small region of the spectrum. The dispersion is not a constant over a single order, and is different for successive orders. An average for each order is used in data reduction, and when orders are articulated to make one long, continuous spectrum, there are small discontinuities in dispersion from order to order. It is possible and practical to interpolate the spectrum in each order and to set all their dispersions equal, but there remains the effect of averaging over each order. There are other, very small errors, rarely observable yet still present, such as "errors of coincidence" that were observed in Harrison's laboratory. When the orders are short (2 nm or less, as an example), there are not many calibration lines in each order, and lines from several orders were used to calibrate the wavelength scale. It was found that lines with the same order–angstrom product did not exactly coincide on the detector (photographic plates), as they theoretically should. It was surmised that the shape of the grooves had an effect in shifting the center of gravity of lines of different wavelengths. No further study was made, except to note this fact in striving for the utmost accuracy in wavelength measurements. Another fact is that the signal-to-noise ratio can change from one end of the order to the other, and judgment is required to extract the best compromise when connecting the orders.

Perhaps we have excessively eulogized BABAR and ignored other, newer instruments (some of which are superior in cardinal parameters — resolution in particular); but we offer a simple challenge to any other instrument: Achieve the theoretical performance in resolution, bandwidth, and signal-to-noise ratio. Simply put, the continued pilgrimage to the mountain to take spectra is the only peer review that counts. Furthermore, when observing (with any instrument), recall "the rules" — especially about the compromise between resolution and noise and the sampling to achieve the radiometric performance necessary for the specific measurement. The performance metrics established in Chapters 7, 8, and 9 for phase and noise performance are difficult to meet with any instrument. The performance nomographs, such as Figs. 2.2 and 5.9, define the optimal strategy for observation and should be used with every instrument and observation.

In addition, BABAR has been cared for and maintained by a dedicated staff who ensure that the instrument is optimally aligned and has maximum modulation efficiency at all times. Consequently we have never really had to deal with misalignment as a regular occurrence; and as laboratory spectroscopists, we would typically realign the interferometer and rerecord a spectrum rather than fuss with an imperfectly observed interferogram. However, over the years we have experimented with the effects of oblique rays, misalignment, and improper (nonuniform) illumination. We have gone round and round about the possible effects of oblique rays as well as of uneven diaphragm irradiation. At one time we tried to find out experimentally the effects of some of these maladjustments. One decision was that the user should not live with misaligned or poorly adjusted instruments. If the effects show up, readjust the instrument until they don't. We tried to correlate off-axis adjustments with particular shapes of spectral lines, but we simply could not identify, let alone correct, the parameters that were important.

As an aside, we note that most of our experience is with a slow instrument — f/35, where misalignment errors are more tolerable. Faster instruments with large fields of view present another class of challenges, and another generation of spectroscopists is actively working on experimental methods for optimizing the performance of such instruments.

Finally, there is a strange no-man's land between instrument design and signal-processing design: Instruments and budgets are inherently finite, and sooner or later there will be opportunities to take data that violate the design specifications of the instrument. A common example is *underresolving* a spectrum when an instrument has insufficient resolution. But phase correction, nonlinearity correction, and speed correction are similarly common examples of second-order effects that plague precision instruments. Each of these instrumental artifacts can be corrected, to varying degrees, with hardware and software solutions. In practice, the software solutions have often been the necessary solution (because the hardware was untouchable) and the practical solution (because they are cheap, effective, flexible, adaptable, and reprogrammable as needed). Of necessity, each of us has *fixed* instruments over the years — typically this is done by systematically decomposing the signal chain (from photon to byte) into characterizable and testable components and building simple models that can be strung together to describe the total system. Ultimately, this process yields a revised definition of the performance specification and the quality (precision and accuracy) of results that can be obtained from the experiment. In Section 11.1, the "Brault algorithm" for correcting speed variations is actually the third "Brault algorithm," following the phase algorithm (Chapter 7) developed in the 1970s and resolution recovery algorithm (Chapter 9) developed in the 1980s. Each algorithm addressed a second-order artifact in the instruments that could be best addressed using a software solution. That the artifact could be addressed and removed with a hardware solution is an irrelevant side note: Given a choice between an absurdly complex servo system and an inexpensive resampling algorithm to correct for speed variations, it is reasonable to favor the latter. The fact that it took 20 years for the *concept* of a better approach to be discovered and demonstrated is an interesting comment on our collective creativity.

As a historical aside, there is a delightful little paper by Filler from 1967 that is focused on analytic continuation — the extrapolation of interferograms to achieve increased resolution. He concludes with a challenge: "The production of a good solution for the extrapolation problem is not merely an intellectual exercise, and it is hoped that one will be forthcoming." Filler's solution would remain hidden for more than 30 years, not unsolved, just undocumented and unpublished. Ironically, in the larger community, when it is impossible to "fully resolve" a spectrum, it is standard practice to apodize the spectrum, destroying the virtues of the FTS ILS function.

Despite the previous assertions of "perfection", there have been cases where the McMath–Pierce 1-m FTS has had inadequate spectral resolution. A specific case in point (illustrated in Chapter 9) is infrared spectra of low-temperature diffusion flames with linewidths of less than 0.015 cm^{-1}, which yields 1.5 samples per FWHM (assuming 0.01 cm^{-1} resolution). A brief examination of Fig. 5.9 will indicate that the distortion of the radiometry is between 4 and 10% — clearly unacceptable. Initially, since we were interested in line positions and equivalent widths, we heavily apodized the spectra and proceeded to reduce the line positions to molecular parameters. The results, as measured by the standard deviation of the molecular analysis, were grossly worse than what we expected based on Eq. (9.2). The solution was obviously to be found in addressing the ILS distortion of the spectrum in a more rigorous fashion. Late one night we discussed the problem and determined that a similar problem had been solved a decade earlier with N_2O calibration spectra — the solution was trivial: Just include the ILS in the fitting process and it will fall out of the spectrum. Due to the fact that the FTS ILS function (to first order) is simply a sharp cutoff filter, it was trivial to code a simple test case, corrupt it with noise, underresolve it, and fit it using a bandwidth-limited version of a least-square fitting algorithm: Magically the ILS distortion melted away with a few dozen iterations. In 30 minutes, two years of work became instantly obsolete; more importantly, another insight into high performance FTS became standard practice. Within months it was coded into our standard fitting algorithms and in use by a majority of the spectroscopists processing FTS data with Brault's codes. Unfortunately, while results from this algorithm became commonplace in the spectroscopic literature starting in 1988, the algorithm and its implications for future instrument design were never published. The paper has been written, the figures and test cases documented, and the authors (who shall remain nameless) have agreed to belatedly publish the work in the refereed literature. As we look back on it today, it seems rather obvious and trivial — even if 13 years later it is still capable of causing quite a stir among instrument builders who do not approve of using software to correct the debits of our instruments.

A word of caution: ILS reconstruction and recovery has nothing to do with various forms of deconvolution. Deconvolution, in most cases, is constrained division by zero, which creates interesting artifacts if used inappropriately. Conversely, multiplication by zero is fully deterministic and cannot introduce computational artifacts. We do not want to give anyone the impression that "an off-line computer will resuscitate what the spectrometer left out," although the things astronomers do with data so as to draw sweeping conclusions often fall well outside what we think is justifiable. But they are often successful. They go where "*spectroscopists* fear to tread." When it is impossible to fully resolve a spectrum, ILS reconstruction offers an alternative to apodization and preserves the information content of the spectrum.

11.6 On Fourier Transform Spectrometry and Digital Signal Processing

There is a latent enigma contained in this text: While the subject and title is *Fourier Transform Spectrometry*, there is an immense amount of digital signal processing both implicit and explicit within the text. This is a natural consequence of the fact that to build an FTS requires a marriage of physical optics (to make the interferogram) and digital signal processing (to demultiplex the interferogram into a spectrum). Historically, this marriage has required two individuals (Janine and Pierre Connes are an obvious example, both literally and metaphorically); in the case of the McMath–Pierce FTS, this marriage occurred in a single individual. When this occurs it is possible to work in both Fourier domains simultaneously — or more typically to flip domains as suits the problem at hand. This skill is most typically taught in electrical engineering, and best exemplified by Bracewell's text *The Fourier Transform and Its Applications*, now in its third edition. The software used to process our data and to generate all of the figures in this text was generated with Brault's *DECOMP* (and its derivatives *GREMLIN* and *GREMLET*), which evolved from a code written to demonstrate the concepts and figures in Bracewell's book. In teaching FTS design it is often necessary to teach a parallel course in digital signal processing to enable full exploitation of instruments and instrument designs.

There is a particularly interesting question as to whether issues of signal processing belong in a book on Fourier transform spectrometry, since they are clearly generic to any type of spectral separation device. Our goal, as spectroscopists and instrument builders, has been to obtain the "best spectra" possible within the constraints of the laws of physics — an admirable goal, but one lacking a quantitative metric without Eq. (9.2). The derivation of a relationship between the precision of a line position measurement and the signal-to-noise ratio and sampling of a spectrum defines the ultimate test of an instrument design. The true magic of the 1-m FTS design was its nearly absolute conformity to theoretical performance.

Historically, the emphasis was on line picking and not data sampling. Typically, users of sampled data do not document the process of picking lines, measuring intensities, assigning line shapes, etc. The data are processed with canned algorithms, and error analysis is typically cursory at best. Reality may be harsh. But how often do we re-examine the source code to understand the assumptions in the measurement process? Various tricks and optical configurations were the rule, as in Harrison's work making the M.I.T Wavelength Tables.

FTS development, in the 1960s and 1970s, was enabled by digital computers and the Fast Fourier Transform, and the same individuals who developed the demultiplexing processes for FTS also developed the data reduction codes that enabled consistent and uniform processing of unprecedented high-resolution broadband spectra. Users of the 1-m FTS went home with digital data tapes and the associated data reduction codes. While many of the algorithms were developed for application to grating spectra, they were directly applicable to FTS data — with a peculiar twist. Common practice in grating spectrometry is to ensure that a minimum of five samples per FWHM are obtained in a spectrum; in contrast, FTS instruments are operated just above Nyquist sampling (two samples per FWHM), to maximize the free spectral range. Extracting precision radiometry from minimally sampled data is difficult at best and requires considerable care in the design of the processing algorithms.

In order to achieve high-precision spectrometry, equal attention needs to be placed on instrument design, operation, and signal processing. Each step in the processing introduces noise into the measurement, and in order to achieve the desired performance, sufficient signal bandwidth and margin must be maintained to ensure a quality measurement.

11.7 From Echelle Spectrographs to Echelle Spectrographs — A 50-Year Circle

The following discussion may appear to be a detour from the subject of FTS and interferometry; to the contrary, in the period between 1911 and 1960 the interferometer was employed to control ruling engines for the manufacture of gratings. Indeed, Michelson's last interferometer became the foundation of the grating industry in the period between 1948 and 1960.

Broadband spectra of modest quality are most simply and cheaply obtained by the grating with photographic or CCD recording, at least in the visible and UV. This system is also the most tolerant of source intensity variation.

Echelle spectrographs with array detectors can be at least an order of magnitude improvement in quantity of data gathered with a diffraction grating and in digital data processing. However, at high-resolution they reproduce line shapes and positions with only modest accuracy, owing to optical aberrations and nonlinearities in dispersion. Data reduction and analysis initially require a minimum of computation to get a first look at the spectrum, but the extra computations required to convert wavelengths to wavenumbers, fit the spectral lines, and construct atlases are time consuming and full of pitfalls. A typical spectrum might consist of 20 successive echelle orders, each with a variable dispersion within an order and a changing dispersion from order to order. The data are in wavelengths rather than wavenumbers, and consequently they require an extra computation to get the energies of levels. The number of samples in each spectral line must be much larger than for FTS data to get accurate fits for position, intensity, shape, width, and area, and even then the lines are always asymmetrical in shape. When constructing atlases, each order must be interpolated to the same dispersion linear in wavenumber, and then the orders must be trimmed and shifted to match each one with the preceding and succeeding ones. These computations are all doable but not trivial.

How did echelles come about? Because they are based on sound optical and data-processing principles and are widely used, a few words about their development might be in order.

At MIT in the 1950s Harrison began ruling *echelles*, a term he coined to indicate a grating with only a few grooves, perhaps 100/mm rather than the 1200/mm commonly used at that time. Some years earlier he had attended a meeting (in the 1930s) to talk about what instrumentation was needed in spectroscopy. At that time he resolved that he would never be trapped into ruling gratings. In 1948, however, a ruling engine at Chicago needed a new home, and he reversed his previous position. He thought an interferometer could be used for precision distance control, and he decided to give grating ruling a try.

He set up the interferometrically controlled ruling engine but soon discovered problems that might make the task an infinite sink for time and effort. The speed of the engine and the resultant speed of ruling were governed by practicalities of mechanical motion. The screw was something like 1 meter long and 10 cm in diameter, and the nut on which the carriage was mounted was still larger. The diamond ruling mechanism was a "four-bar system" well known to mechanical engineers, with bars on the order of 1 meter long. They were adjusted to minimize curving in the grooves at the ends, where the diamond had to slow down and reverse direction. The rulings were straight otherwise, and of course tilted with respect to the axis of the engine, although only slightly and not enough to cause any problems when mounting the grating. A few millimeters at the ends of the grooves had to be blocked off before the grating could be used.

The rate of ruling was 12 grooves/minute, limited by mechanical motion. The time over which the engine could be made stable was limited to a few days at most. It was in an insulated room, isolated by an anteroom and doors, to keep the temperature constant. No one went into the engine room when the engine was operating. These conditions severely limited the total number of grooves that could be ruled at one time. In principle and in practice, it is no more difficult to rule 10,000 gr/mm than it is to rule 100 gr/mm. It's these stability conditions that are the limitation. Harrison was shooting for a 25-cm-wide grating. To achieve this, he had to restrict the groove density to between 20 and 100/mm. This brought on other problems.

These gratings have grooves ruled by a boat-shaped diamond stylus. They are ruled in an aluminum coating on a glass substrate and have a sawtooth cross section with an angle of 90 degrees, depending how the grating is to be used. The depth of the grooves is on the order of the groove spacing. A 100-gr/mm grating requires a coating depth of about 10 microns, a factor of 10 greater than any coatings of that era. One micron, yes, but thicker than that required so much material that it was not possible to produce a plane surface or one free from aluminum crystals on the surface. Eventually these problems were solved, but only after several years of research and experimentation at Bausch and Lomb.

Part of Harrison's genius was to accept these restrictions and limitations and to design an optical system to utilize what he could produce. In his planning, he recalled some optical properties of gratings that most people ignore. What's important?

1. *Resolving power* $R = \lambda/\delta\lambda$. The resolving power depends on the maximum path difference between rays from the ends of the grating. The grating equation is $m\lambda = d(\sin\theta + \sin\phi)$. In a Czerny–Turner mounting, where the angles of incidence and reflection are approximately the same, with a grating of width W, the path difference is $2W \sin\theta$ and the resolving power is $2W \sin\theta/\lambda$ — the number of wavelengths that will fit into the maximum path difference. *The resolving power is not a function of groove spacing!* A 10-cm-wide grating with 2 grooves or 20 000 grooves has the same resolution, so long as it is used at the same angle.

The resolving limit is the smallest wavelength separation that can be resolved into two spectrum lines, and is given by

$$\delta\lambda = \lambda/R = \lambda^2/2W \sin\theta. \tag{11.16}$$

In wavenumber space the equivalent quantities are

$$R = \sigma/\delta\sigma = 2\sigma W \sin\theta \tag{11.17}$$

and

$$\delta\sigma = 1/2W \sin\theta. \tag{11.18}$$

2. *Reciprocal dispersion or plate factor* $d\lambda/d\theta$. Again,

$$m\lambda = d(\sin\theta + \sin\phi) \tag{11.19}$$

and

$$(d\lambda/d\theta) = \lambda/2\tan\theta. \tag{11.20}$$

The spectrum is spread out more when θ gets larger and spreads rapidly as θ gets larger than 60 degrees.

3. *Free spectral range* F_λ, or F_σ. Two different wavelengths appear at the same angle when $m\lambda_1 = (m+1)\lambda_2$, or $(\lambda_1 - \lambda_2) = \lambda_2/m$. For the case where the two wavelengths are nearly the same and m is large,

$$F_\lambda = \delta\lambda = \lambda/m = \lambda^2/2d\sin\theta \tag{11.21}$$

and correspondingly

$$F_\sigma = 1/2d\sin\theta. \tag{11.22}$$

4. *Blaze, blaze angle, and free spectral range*. When discussing diffraction gratings, elementary texts treat a transmission grating, a plane opaque screen with a number of parallel slits of width a and spacing d through which light is transmitted and diffracted. The grating equation gives the positions of interference maxima. The zeroth-order maximum appears at $\theta = 0$ and higher orders at angles given by

$$\sin\theta = +/-m\lambda/d. \tag{11.23}$$

When only interference is considered, all orders have equal intensities. In fact, interference maxima are multiplied by the envelope of a single slit diffraction pattern. This envelope has maximum intensity in the zeroth order. The zeroes of intensity fall at $\sin\theta = n\lambda/a$, where a is the slit width. Light from the source, striking the grating, is spread out into many orders of decreasing intensities as the order numbers get larger. How many? For a given wavelength, the maximum value of m is d/λ, so there are nominally $2d/\lambda + 1$ orders. The resolution and dispersion get larger as the angle gets larger, but the intensity of the orders gets smaller as the angle gets larger. A given order has very little of the total light intensity.

The question then arises, how can we channel the spectral energy into a single order, or at most a very few, so we can make use of the dispersed light? One way is to decrease the slit size until the central diffraction maximum gets very wide, never reaching zero even with a diffraction angle of 90 degrees. The problem here is that the total amount of light passing through the slits is small.

On the other hand, the slits could be made so wide that the central diffraction maximum becomes so narrow that it covers only one or at most two interference maxima. Another way to look at this is to say that the central single slit diffraction maximum covers one or at most two free spectral ranges. All the energy goes into one or two orders only, for all practical purposes. For example, suppose we could make the slit width the same as the slit separation, $a = d$. Then the diffraction envelope would have a maximum at $\theta = 0$ and zeroes at $\sin\theta = n\lambda/d$, with $n > 0$. Interference maxima would occur at $\sin\theta = m\lambda/d$. All of the light would fall under the central diffraction maximum. However, this occurs at $\theta = 0$, so we get no useful information from the multiple slit interference. Going a step further, is there any way to move the maximum of the single slit pattern so that it is centered on an order other than the zeroth one? The answer is yes, and the optics are more easily understood when we use a reflection grating.

A reflection grating is one in which the grooves (slits) reflect light rather than transmitting it. A flat surface with highly reflecting strips would duplicate the preceding considerations with regard to a transmission grating. But consider another arrangement. We can think of a reflection grating as being like a stair step, where each groove has a "rise" and a "run" that are perpendicular to one another. Let the light be incident parallel to the run, perpendicular to the rise, and reflected directly backward (or almost) so that the reflection is specular and $\theta = \phi$, with the groove cross section chosen so that these angles are 45 degrees with respect to the surface of the grating. In this case, with a grating spacing of d, the rise is $0.7d$, the run is $0.7d$. The path difference between adjacent beams is twice the run. Now the maximum of the single "slit" diffraction pattern is at $\theta = \phi = 45$ degrees, while the order of interference is $2(.7)d/\lambda$. The zeroth order of single-slit diffraction coincides with the mth order of interference, where the mth order is given by $m = (d/\lambda)\sin\theta$. In this case the "slit" width is not equal to the groove spacing, so the central diffraction maximum covers more than a single interference order, by a factor of 1/.7 or 1.4. And if the interference pattern were not a maximum exactly at the proper angle, there would be perhaps two more orders. It is still true that a great fraction of the light is diffracted at the desired angle, and the grating is said to have a "blaze", or to be blazed at the angle specified or at the central wavelengh in the first order.

An echelle spectrometer or spectrograph must include a second low-dispersion element, to produce cross dispersion to separate the orders. The cross-dispering element can be inside or outside the instrument (the latter only if the spectrograph is stigmatic (a point on the slit goes into a point on the detector — a photographic plate). Much of the early work showing the feasibility of an echelle spectrograph was done with a 20-gr/mm echelle and a concave grating used as an order sorter consisting of a concave grating in place of one of the mirrors. Each order was about 50 mm in length. The cross-dispersion was large enough to have a spectrometer entrance slit width of about 3 mm in length. The photographic plates using for recording the spectrum were 50 mm $\times$ 460 mm, as we commonly used with a 6.4-meter concave grating.

In later instruments, we installed a prism spectrometer outside the Czerny–Turner mounting and focused its output spectrum on the entrance of the CT spectrometer. We adjusted the entrance slit on the prism spectrometer to cover only a single free spectral range of the echelle. To illlustrate with an example: Echelle with a width of 10 cm, 75-gr/mm, blaze angle 63.4 degrees (tan θ = 2) m = 60 for 500 nm, resolving power 230,000, resolving limits 0.02 Å and 0.08 cm^{-1}, reciprocal dispersion 1250 Å/radian and 5000 cm^{-1}/radian, and a free spectral range of 83 Å and 330 cm^{-1}. For a 1-meter focal length instrument, the angular spread for one free spectral range covers 66 mm. A suitable cross-disperser, of whatever design, will be needed to separate these orders.

Early on there was much discussion of standardizing echelle parameters so a manufacturer or an individual spectroscopist could have a reasonably standardized model to shoot for. Harrison favored simplicity and high angles of incidence to take advantage of resolution and dispersion, and decided to make all groove shapes with run/rise = 2, making tan θ = 2, and θ = 64 degrees.

Suppose you are a new user of an echelle. What features (bugs, too) would strike you?

1. The instrument is relatively compact; no long focal lengths are necessary.

2. The cross-dispersion makes the instrument more complex, and more difficult keep in adjustment.

3. The spectrum is never truly in good focus. Focus is always a distinct mpromise because there are off-axis optical elements in two dimensions (skew /s).

4. The spectral display is freaky because it is two dimensional. A visual display th vertical spectral lines moves up or down as the wavelength region changes — spectral lines are not perpendicular to the baseline of the spectrum. It takes me psychological reorientation to get used to this.

5. The spectral lines are short so as to avoid overlapping.

6. The dispersion across a single order is not a constant. Remember, the persion goes as the tangent of the angle, and the rate of change at high angles is ge.

7. Adjacent orders will cover several of the same lines. When piecing the ctrum together to make a single long one for display and computation, a decision to be made as to where to make the cutoff in one order and to start in the next ler. Invariably the two orders, where they overlap, are not the same in dispersion, ensity, sharpness, and signal-to-noise ratio. It takes a lot of judgment and hand rk to do this assembling of the spectrum.

8. It's great to have the entire spectrum in a compact form. You can see the ole spectrum all at the same time and can often pick out regularities that are der to correlate on one long spectrum.

8 A Final Story

Sometimes the *process* of doing experimental research loses something in nslation. Figure 11.2 illustrates a surprisingly common exercise — trying to ke a King furnace (a high-temperature furnace used to simulate cool carbon rs) behave. This is not a staged picture: The hammer is real and being used gressively to dislodge filamentary structures growing on the interior walls of the nace. When the filaments become sufficiently large, they break off the walls l fall through the FTS field of view. The incandesent particles moving in the d of view create horrible spikes in the interferogram and redefine *deglitching* l *baseline variation* in FTS. The ambition of the *hammering* exercise is to clean interior sufficiently to obtain one good interferogram. Years later, the lengths to which one would go to obtain data does not seem all that surprising. That the experiment did not work was irrelevant.

Fig. 11.2 Davis (left) and Brault (right) "deglitching" the King furnace at the 1-m FTS at the McMath–Pierce Solar Telescope.

Interferometry is the art of the second order — carefully managing optical and signal effects typically ignored in imaging systems; once mastered, they offer a new insight into each experiment and indeed the universe. As in Frost's "The road less travelled", this adventure in physical optics and signal processing has provided each of us with the delightful ability to appreciate serendipitous opportunities and to explore them — and in exploring them to be fully satisified, perhaps telling a story about the adventure and perhaps not (if something more interesting comes along). As with a long walk in the woods, or an unsuccessful experiment, one is better for the journey. But it is likely no one else will ever know that the opportunity existed, and unknowingly we are all the worse for it.

12

CHAPTER-BY-CHAPTER BIBLIOGRAPHY

Chapter 1. Introduction

Reviews

Jacquinot, P., New developments in interference spectroscopy, *Reports on Progress in Physics*, **XXIII**, 295–306, 1960. The proceedings of a meeting held in 1957 in Bellevue, France, which provide a complete and excellent review of the subject, summarizing the status of various developments at the time and giving references.

Connes, P., Astronomical Fourier spectroscopy, *Ann. Rev. Astron. Astrophys.*, **8**, 209, 1970. A good early review of the role of Fourier spectrometry in astronomy.

Ridgway, S. T., and Brault, J. W., Astronomical Fourier transform spectroscopy revisited, *Ann. Rev. Astron. Astrophys.*, **22**, 291, 1984. A review that includes references to other reviews as well as many original papers.

Brault, J. W., Fourier transform spectrometry, in *High Resolution in Astronomy*, Proceedings of the Fifteenth Advanced Course of the Swiss Society of Astronomy and Astrophysics, A. O. Benz, M. C. E. Huber, and M. Mayor, eds., Saas Fee, 1985 (Sauverny, Observatoire de Genève, Switzerland), pp. 1–61. Often referred to as the "Saas Fee" notes. They represent the seminal publication on which this book is based.

Introductory Articles

Two excellent introductory articles about Fourier transform spectrometry have been published in *Analytical Chemistry*:

Faires, L. M., Fourier transforms for analytical atomic spectroscopy, *Anal. Chem.*, **58**, 1023A, 1986.

Thorne, A. P., Fourier transform spectrometry in the ultraviolet, *Anal. Chem.*, **63**, 57A, 1991.

General Texts

Four informative general texts can be recommended.

Steel, W. H., *Interferometry*, Cambridge University Press, 1983. An overview of interferometry (not just for FTS) by one of the pioneers.

Chamberlain, J., *The Principles of Interferometric Spectroscopy*, Wiley, 1979. Fourier transform spectrometry from a spectroscopist's point of view.

Thorne, A. P., *Spectrophysics*, 2nd ed., Chapman and Hall, New York, 1988.

Thorne, A. P., Litzen, U., and Johansson, S., *Spectrophysics: Principles and Applications*, Springer-Verlag, New York, 1999.

Remote Sensing

In application to remote sensing there are several excellent references:

Hanel, R. A., Conrath, B. J., Jennings, D. E., and Samuelson, R. E., *Exploration of the Solar System by Infrared Remote Sensing*, Cambridge University Press, Cambridge, England, 1992. The grand tour of infrared remote sensing by members of the team that sent FTS instruments to Earth, Mars, Jupiter, Saturn, Uranus, and Neptune and numerous satellites in the process.

Beer, R., *Remote Sensing by Fourier Transform Spectrometry*, Wiley, New York, 1992.

Persky, M. J., A review of spaceborne infrared Fourier transform spectrometers for remote sensing, *Rev. Sci. Instrum.*, **66**, 10, 4763–4797, 1995.

Chapter 2. Why Choose A Fourier Transform Spectrometer?

Fabry, C., and Perot, A., Théorie et application d'une nouvelle méthode de spectroscopie interférentielle, *Annales de Chimie et de Physique*, **16**, 115-244, 1899.

Davis, S. P., *Diffraction Grating Spectrographs*, Holt, Rinehart, and Winston, New York, 1970.

Hanel, R. A., Recent advances in satellite radiation measurements, in *Advances in Geophysics*, H. E. Landsberg and J. Van Mieghem eds., Academic Press, New York, 1970.

Brault, J. W., Fourier transform spectrometry in relation to other passive spectrometers, *Phil. Trans. R. Soc. Lond*, A **307**, 503, 1982.

Vaughan, M., *The Fabry–Perot Interferometer*, Institute of Physics, New York, 1989.

Chapter 3. Theory of the Ideal Instrument

Jacquinot, P., and Dufour, C., Condition optique d'emploi des cellules photo-électriques dans les spectrographes et les interféromètres, *Journal Recherche du Centre National Recherche Scientifique Laboratorie Bellevue (Paris)*, **6**, 91–103, 1948.

Fellgett, P. B., *The Multiplex Advantage*, Ph.D. Thesis, University of Cambridge, Cambridge, U.K., 1951.

Fellgett, P. B., Spectrométre interférentiel multiplex pour mesures infra-rouge sur les étoiles, *Journal de Physique et le Radium*, **19**, 237–240, 1958.

Connes, J., Recherches sur la spectroscopie par transfomation de Fourier, *Revue d'Optique*, **40**, 45–78, 116–140, 171–190, 231–265, 1961. Translated into english by the U.S. Navy (NAV-WEPS Rept. No. 8099, NOTS TP 3157, U.S. Naval Ordnance Test Station, China Lake, CA).

Fellgett, P. B., The origins and logic of multiplex, Fourier, and interferometric methods in spectrometry, *Aspen International Conference on Fourier Spectroscopy*, G. A. Vanesse, A. T. Stair, and D. J. Baker eds., pp. 139–142, U.S. Air Force Cambridge Research Laboratory 71-0019, 1971.

Chapter 4. Fourier Analysis

Mathematical Reference

The following classical reference text continues to be the standard mathematical reference at both the introductory and advanced levels. The pictorial dictionary is remarkable and applicable to many problems.

Bracewell, R. N., *The Fourier Transform and Its Applications*, McGraw-Hill, New York, 1965.

Statistical Signal Processing

Elements of statistical signal processing have been elegantly presented in the following references.

Wiener, N., *The Extrapolation, Interpolation, and Smoothing of Statistical Time Series*, Wiley, New York, 1949.

Lee, Y. W., *Statistical Theory of Communications*, Wiley, New York, 1960.

Direct applications of optimum filtering to cross-correlation are difficult to extract from the classical texts. The Appendix to the text by Beer discusses one application, the definition of the optimum resolution.

Beer, R., *Remote Sensing by Fourier Transform Spectrometry*, Wiley, New York, 1992.

Nonlinearity

Nonlinearity remains a sticky subject where many of the assumptions fail and the sampling becomes insufficient to properly diagnose the performance of the interferometer. Guelachvili presents an experimental method that minimizes the effect, while Abrams *et al.* describes a computational method for removing the effect.

Guelachvili, G., Distortion-free interferograms in Fourier transform spectroscopy with nonlinear detectors, *Appl. Opt.*, **25**, 4644–4648, 1986. A method of reducing the nonlinear effects by using the two outputs of a FTS instrument (analogous to the virtues of a "push–pull" method to minimize nonlinearities in a power amplifier).

Abrams, M. C., Toon, G. C., and Schindler, R. A., Practical example of the correction of Fourier-transform spectra for detector nonlinearity, *Appl. Opt.*, **33**, 27, 6307–6314, 1994. N. B.: Subsequent to publication we determined that the *nonlinearity* attributed to the MCT detector was actually introduced by the amplifiers. Removal of the amplifiers essentially eliminated the nonlinearity and did not decrease the signal-to-noise ratio (MCA).

Chapter 5. Nonideal (Real–World) Interferograms

Connes, J., Domaine d'utilisation de la méthode par transformée de Fourier, *J. Physique et le Radium*, **19**, 197–208, 1958. A detailed discussion of the shape of the instrumental profile, including the effects of finite path length (L) and of the finite field of view (typically a circular, centered diaphragm, optimum diaphragm).

Connes, J., *Recherches sur la spectroscopie par transfomation de Fourier*, Ph.D. Thesis, Laboratory Aime Cotton, Bellevue, France, 1960.

Connes, J., Recherches sur la spectroscopie par transfomation de Fourier, *Revue d'Optique*, **40**, 45–78, 116–140, 171–190, 231–265, 1961. Published version of J.

Connes' thesis presenting the properties of FTS: (a) the theoretical instrumental line shape profile, (b) real instrumental line shape functions, (c) noise in FTS, (d) noise in numerical Fourier transformations, (e) experimental results (night sky spectra). Translated into english by the U.S. Navy (NAV-WEPS Rept. No. 8099, NOTS TP 3157, published by the U.S. Naval Ordnance Test Station, China Lake, CA).

Connes, J., Computing problems in Fourier spectroscopy, Aspen International Conference on Fourier Spectroscopy, G. A. Vanasse, A. T. Stair, and D. J. Baker eds., pp. 83–115 U.S. Air Force Cambridge Research Laboratory, 71-0019, 1971.

Brault, J. W., Fourier transform spectrometry, in *High Resolution in Astronomy: Proceedings of the Fifteenth Advanced Course of the Swiss Society of Astronomy and Astrophysics*, A. O. Benz, M. C. E. Huber, and M. Mayor, eds., Saas Fee, 1985 (Sauverny, Observatoire de Genève, Switzerland), pp. 1–61.

Abrams, M. C., High-precision Fourier transform spectroscopy of atmospheric molecules, Ph.D. thesis, University of California at Berkeley, Berkeley, CA, April 1990. A stepping stone between the "Saas Fee Notes" and this book, with a detailed discussion of resolution enhancement of FTS spectra correcting for distortion of the spectrum by the instrument line shape function.

Genest, J., and Tremblay, P., Instrument line shape of Fourier transform spectrometers: analytic solutions for nonuniformly illuminated off-axis detectors, *Appl. Opt.*, **38**, 5438–5446, 1999.

Chapter 6. Working with Digital Interferograms, Fourier Transforms and Spectra

Rayleigh, Lord, *Phil. Mag.*, **8**, 5, 261, 1879.

Nyquist, H., Certain topics in telegraph transmission theory, *A.I.E.E. Trans.*, **617**, 644, 1928.

Blackman, R. B., and Tukey, J. W., *The Measurement of Power Spectra*, Dover, New York, 1958.

Bracewell, R. N., *The Fourier Transform and Its Applications*, McGraw-Hill, New York, 1965.

Cooley, J. W., and Tukey, J. W., *Mathematics of Computation*, **19**, 296, 1965. Introduction of the fast Fourier transform (Cooley–Tukey algorithm).

Forman, M. L., Fast Fourier transform technique and its application to Fourier spectroscopy, *J. Opt. Soc. Amer.*, **56** 978–979, 1966.

Cochran, W. T., Cooley, J. W., Favin. D. L., Helms, H. D., Kaenel, R. A., Lang. W. W., Maling, G. C., Nelson. D. E., Rader, C. M., Welch, P. D., *IEEE Trans. Audio Electroacoustics, Special Issue on Fast Fourier Transform and Its Application to Digital Filtering and Spectral Analysis*, AU-15, **2**, 45, 1967. A complete discussion of FFT algorithms and applications — the definitive reference.

Brault, J. W., and White, O. R., The analysis and restoration of astronomical data via the fast Fourier transform, *Astron. Astrophys.*, **13**, 169–189, 1971. A thorough introduction to optimum filtering for the restoration of spectra using the FFT. As a historical note, this paper precedes Brault's work in FTS and is focused on the restoration of grating spectra.

Rabiner, J., ed., *Programs for Digital Signal Processing*, IEEE Press, New York, 1972.

Bracewell, R. N., Numerical transforms, *Science*, **248**, 647–704, 1990.

Apodization

Apodization is a sticky subject with spectroscopists. It was introduced in the 1960s when computational resources were very limited. By the 1980s it was largely unnecessary, but nevertheless it continues to be used by many groups. As illustrated in Chapter 6, it is necessary to decrease the resolution of a spectrum by more than 50% in order to suppress the instrumental line shape (sinc function) of the FTS when there is no self-apodization. Fellgett (1967) asserted that *apodization should be done only by experts*. We do not advocate the usage of apodization in routine processing of spectral data. In its place we describe in Chapters 6 and 9 methods for fitting properly the instrument function and recovering the full resolution. If interferograms have been saved it is possible to reprocess historical data and recover the full performance of the instrument.

Jacquinot, P., Quelques recherches sur les raise faibles dans les spectres optiques, *Proc. Phys. Soc.*, **63**, 12, 969–979, 1950.

Dossier, B. *Rev. Opt.*, **33**, 57, 1954. The first suggestion to change an instrumental profile by suitable screening of a grating (apodization before the word was invented).

Gebbie, H. A., *J. Phys. Radium*, **19** 230, 1958. First use of apodization in computational processing.

Filler, A. H., Apodization and interpolation in Fourier-transform spectroscopy, *J. Opt. Soc. Amer.*, **54**, 762, 1964. A concise treatise on apodization in which the concept of a *Filler diagram* (mapping sidelobe suppression *vs.* distortion of the central lobe) was introduced.

Fellgett, P., Interventions regarding a paper by Beer and Cayford, *J. Physique*, **28**, C2-38–39, 1967.

Norton, R. H., and Beer, R., New apodizing functions for Fourier spectrometry, *J. Opt. Soc. Amer.*, **66**, 259–264, 1976; **67**, 419, 1977. The erratum in volume **67** is rarely referenced.

Harris, F. J., On the use of windows for harmonic analysis with the discrete Fourier transform, *Proc. IEEE*, **66**, 51–83, 1978. A vast expansion of the work done by A. H. Filler with the introduction of a signal-to-noise gain metric. Apodization windows include: rectangle, triangle, Hanning ($\cos^2$), Hamming, Rietz, Riemann, de la Valle–Poussin, Tukey, Bohman, Poisson, Hanning–Poisson, Cauchy, Gaussian, Dolph–Chebyshev, Kaiser–Bessel, Barcilon–Temes, exact Blackman, and Blackman–Harris.

Chapter 7. Phase Corrections and Their Significance

Mertz, L., *Transformations in Optics*, Wiley, New York, 1965.

Forman, M. L., Steel, W. H., and Vanasse, G. A., Correction of asymmetric interferograms obtained in Fourier spectroscopy, *J. Opt. Soc. Amer.*, **56**, 59–63, 1966.

Mertz, L., Auxiliary computation for Fourier spectroscopy, *Infrared Phys.*, **7**, 17, 1967.

Connes, J., Computing problems in Fourier spectroscopy, Aspen International Conference on Fourier Spectroscopy, G. A. Vanasse, A. T. Stair, and D. J. Baker eds., pp. 83–115, U.S. Air Force Cambridge Research Laboratory, 71-0019, 1971.

Mertz, L., Fourier spectroscopy, past, present, and future, *Appl. Opt.*, **10**, 386–389, 1971.

Brault, J. W., High-precision Fourier transform spectrometry: the critical role of phase corrections, *Mikrochim. Acta [Wien]*, **III**, 215, 1987.

Learner, R. C. M., Thorne, A. P., Wynne-Jones, I., Brault, J. W., and Abrams, M. C., Phase correction of emission line Fourier transform spectra, *J. Opt. Soc. Amer. A*, **12**, 10, 2165–2171, 1995.

Chapter 8. Effects of Noise in Its Various Forms

Birk, M., and Brault, J. W., Detector quantum efficiency: an important parameter for FT-IR spectroscopy, *Mikrochim. Acta [Wien]*, **II**, 243–247, 1988.

Learner, R. C. M., Thorne, A. P., Brault, J. W., Ghosts and artifacts in Fourier-transform spectrometry, *Appl. Opt.*, **35**, 2947–2954, 1996.

Chapter 9. Line Positions, Line Profiles, and Line Fitting

Harrison, G. R., and Molnar, J. P., *J. Opt. Soc. A*, **30**, 343, 1940.

Savitsky, A., and Golay, M. J. E., Smoothing and differentiation of data by simplified least squares procedures, *Anal. Chem.*, **36**, 1627, 1964. One of the few historical references on *line finding* by differentiation. The procedures and conclusions are not numerically optimal.

Ziessov, D., *On-line Rechner in der Chemie. Grundlagen und Anwendung in der Fourier Spektroskopie*, Walter de Gruyter, Berlin, pp. 345–354, 1973. An considerable expansion of the work of Savitisky and Golay with numerous corrections.

Connes, J., Computing problems in Fourier spectroscopy, Aspen International Conference on Fourier Spectroscopy, G. A. Vanasse, A. T. Stair, and D. J. Baker eds., pp. 83–115, U.S. Air Force Cambridge Research Laboratory, 71-0019, 1971.

Voigtian Function

The voigtian function is the convolution of a lorentzian line profile function with a gaussian line profile, first proposed by Voigt , and developed into analytic functions suitable for numerical computation of line shapes by Van de Hulst and Reesinck. Tables of numerical voigtian functions have been derived by Elste.

Van de Hulst, H. C., and Reesinck, J. J., *Astrophysic. J.*, **106**, 121, 1947.

Elste, G., Die Entzerrung von Spektrallinien unter Verwendung von Voigtfunktionen, *Z. für Astrophysik*, **33**, 39–73, 1953.

Abrams, M. C., High-precision Fourier transform spectroscopy of atmospheric molecules, Ph.D. Thesis, University of California at Berkeley, Berkeley, CA, April 1990. A discussion of resolution enhancement of FTS spectra by correcting for distortion of the spectrum by the instrument line shape function. The results presented here represent a recalculation of the results of Elste suitable for full single-precision numerical computation of voigtian functions. The integrations were performed by Brault and are an integral part of the process of fitting line profiles in *GREMLIN*.

Chapter 10. Processing of Spectral Data

Davis, S. P., Laboratory spectroscopy of astrophysically interesting molecules, *Proc. Astron. Soc. Pac.*, **99**, 1105, 1987.

Davis, S. P., and Littleton, J. E., Line intensity determinations for the Red System of CaH, unpublished, 1989.

Chapter 11. Discussions, Interventions, Digressions, and Obscurations

Apodization

Dolph, C. L., A Current Distribution for Broadside Arrays Which Optimizes the Relationship between Beam Width and Side-Lobe Level, *Proc. I.R.E.*, **34**, 335–348, 1946.

Jacquinot, P., Quelques recherches sur les raise faibles dans les spectres optiques, *Proc. Phys. Soc.*, **63**, 12, 969–979, 1950.

Bartlett, M. S., Periodogram Analysis and Continuous Spectra, *Biometrika*, **37**, 1–16, 1950.

Jacquinot, P, The luminosity of spectrometers with prisms, gratings, or Fabry-Perot etalons, *J. Opt. Soc. Amer.* 44, 761, 1954.

Filler, A. H., Apodization and interpolation in Fourier transform spectroscopy, *J. Opt. Soc. Amer.*, **54**, 762–767, 1964.

Filler, A. H., Some mathematical manipulations of interferograms, *J. Physique*, 28, C2–14, 1967.

Beer, R., and Cayford, A. H., An investigation of a fundamental intensity error in Fourier spectroscopy, *J. Physique*, **28**, C2-38–39, 1967.

Helms, H. D., Digital filters with equiripple or minimax responses, *IEEE Trans. Audio Electroacoust.*, **AU-19**, 87–94, 1971.

Norton, R. H., and Beer, R., New apodizing functions for Fourier spectrometry, *J. Opt. Soc. Amer.*, **66**, 259–264 (1976); Erratum *J. Opt. Soc. Amer.*, **67**, 419, 1977.

Harris, F. J., On the Use of Windows for Harmonic Analysis with the Discrete Fourier Transform, *Proc. I.E.E.E.*, **66**, 51–83, 1978.

Lee, J. P. and Comisarow, M. B., Apodization Functions for Absorption-Mode Fourier Transform Spectroscopy, *Appl. Spec.*, **43**, 599–604, 1989.

Imaging Fourier Transform Spectrometers (IFTS)

Bennett, C. L., Fourier transform IR measurements of thermal infrared sky radiance and Transmission, *Proc. SPIE*, 2266, 25–35, 1994.

Bennett, C. L., Effect of jitter on an imaging FTIR spectrometer, *Proc. SPIE*, 3063, 174–184, 1997.

Bennett, C. L., Carter, M. R., Fields, D. J., Hyperspectral imaging in the infrared using LIFTIRS, *Proc. SPIE*, 2552, 274–283, 1995.

Bennett, C. L., Carter, M. R., Fields, D. J., Lee, F. D., Infrared hyperspectral imaging results from vapor plume experiments, *Proc. SPIE*, 2480, 435–444, 1995.

Bennett, C. L., Carter, M. R., Fields, D. J., Hernandez, J. A., Imaging Fourier transform spectrometer, *Proc. SPIE*, 1937, 191–200, 1993.

Carter, M. R., Bennett, C. L., Fields, D. J., Hernandez, J. A., Gaseous effluent monitoring and identification using an imaging Fourier transform spectrometer, *Proc. SPIE*, 2092, 16–26, 1994.

Graham, J. R., Abrams, M. C., Bennett, C. L., Carr, J., Cook, K., Dey, A., Wishnow, E., The performance and scientific rational for an IR imaging Fourier transform spectrograph on a large space telescope, *Proc. Astro. Soc. Pac.*, **110**, 1205, 1998.

Maillard, J.-P., 3D Spectroscopy with a Fourier transform spectrometer, in *3D Optical Spectroscopic Methods in Astronomy*, ASP Conference Series, G. Comte and M. Marcelin eds., **71**, 316, 1995.

13

CHRONOLOGICAL BIBLIOGRAPHY

This bibliography text was started by L. Delbouille and G. Roland. Their contributions are gratefully acknowledged.

Prehistory: 1881–1950

Michelson performed early measurements on the speed of light with amazing skill and in 1881 perfected the interferometer that bears his name. His purpose was to discover the effect of the Earth's motion on the observed speed (Michelson, 1881, Michelson and Morley, 1887). The result, that light travels at a constant speed, set the stage for the theory of special relativity. In addition, the interferometer was used by Michelson to find suitable spectral lines for metrology systems — his measurement of the standard meter in terms of wavelengths of cadmium light remained the standard for nearly 70 years. Spectral information was obtained by visually recording the "visibility curve" (literally using his eye as the detector) as the interferometer was stepped. Simple spectra were obtained by applying a Fourier transform and enabled the classification of various emission lines as singlet, doublet (yellow sodium line), and multiple (green line of mercury). Michelson and Stratton (1898) constructed a mechanical harmonic analyzer with which Michelson studied the broadening of spectral lines. Rubens and Wood (1911) recorded the first interferogram and performed spectral analysis by guessing the number and strength of the components within a narrow spectral range and synthesizing interferograms.

Michelson, A. A., The relative motion of the Earth and the luminiferous ether, *Amer. J. Sci.*, **22**, 120–129, 1881.

Michelson, A. A., and Morley, E. W., On the relative motion of the Earth and the luminiferous ether, *Amer. J. Sci.*, **34**, 333–345, 1887.

Michelson, A. A., and Stratton, S. W., A new harmonic analyzer, *Amer. J. Sci.*, **5**, 1–13, 1898.

Rubens, H., and Wood, R. W., *Phil. Mag.*, **21**, 249, 1911.

Michelson, A. A., *Studies in Optics*, Phoenix Books, University of Chicago Press, Chicago, 1927. In particular see the chapter entitled "Visibility of Fringes."

Wood, R. W., *Physical Optics*, 3rd ed., McMillan, New York, 1934.

Development of FTS: 1945–1970

In the years immediately following World War II two researchers realized the potential of Michelson interferometers as spectrometers with specific performance "advantages" relative to scanning grating spectrometers. Jacquinot and Dufour (1948) recognized the throughput advantage of a spectrometer with a circular aperture rather than a narrow slit. Termed the *Jacquinot* or *étendue advantage*, this energy-gathering capability has been fully exploited by laboratory and space-based interferometers. During the same period of time, Fellgett (1951) recognized the "multiplex" advantage of a two-beam interferometer, which can measure all spectral intervals simultaneously instead of sequentially as in a scanning grating spectrometer. For equal observing time, the signal-to-noise ratio improves by the square root of the number of spectral intervals for source-noise-limited observations. The final necessary element was the exploitation of the digital computer for calculating the Fourier transform of the interferogram, first suggested in a paper on the lamellar grating instrument (Gebbie and Vanasse, 1956).

Jacquinot, P., and Dufour, C., Condition optique d'emploi des cellules photoélectrique dans les spectrographes et les interféromètres, *Journal Recherche du Centre National Recherche Scientifique Laboratorie Bellevue (Paris)*, **6**, 1, 1948.

Fellgett, P. B., *The Multiplex Advantage* Ph.D. Thesis, University of Cambridge, Cambridge, U.K., 1951.

Fellgett, P. B., *J. Opt. Soc. Amer.*, **42**, 872, 1952a.

Fellgett, P. B., Symposium on Molecular Structure and Spectroscopy, Ohio State University, Columbus, OH, 1952b.

Jacquinot, P., *The Etendue Advantage*, XVII me Congrs du GAMS, Paris, 1954.

Jacquinot, P., *J. Phys. Radium*, **19**, 223, 1958.

Fellgett, P. B., Spectrométre inferférentiel multiplex pour mesures infra-rouges sur les étoiles, *Journal de Physique et le Radium*, **19**, 237–240, 1958.

First experimental demonstration of the multiplex advantage on stellar spectra. N.B.: J. Connes and P. Connes note in their 1966 paper that Fellgett "emphasized that *multiplex* spectrometry can be realized in other ways than interferometrically, and that orthogonal functions other than the trigonometrical ones can be used. Furthermore, multiplexing is a property of the observing system as a whole. The use of interferometry or of Fourier methods or of both does not guarantee that useful multiplexing of the spectral elements will be realized."

Fellgett, P. B., The origins and logic of multiplex, Fourier, and interferometric methods in spectrometry, Aspen International Conference on Fourier Spectroscopy, G. A. Vanasse, A. T. Stair, and D. J. Baker eds., pp. 139–142, U.S. Air Force Cambridge Research Laboratory, 71-0019, 83–115, 1971.

Gebbie, H. A., and Vanasse, G., *Nature*, **178**, 432, 1956.

Gebbie, H. A., Vanasse G., and Strong, J., *J. Opt. Soc. Amer.*, **46**, 377, 1956.

Gebbie, H. A., *Phys. Rev.*, **107**, 1194, 1957.

Strong, J., *J. Opt. Soc. Amer.*, **47**, 354, 1957.

The first significant discussion of the new field of *Fourier transform spectroscopy* occurred at the "Colloque International sur les Progrès Récents en Spectroscopie Interfrentielle," Bellevue, France, Sept. 9–13, 1957. Eight papers, out of a total of 51, were devoted to FTS. Attended by the pioneers of FTS, this meeting was the first occasion for a discussion of the merits of FTS. The results were published in *Revue de Physique et le Radium* in 1958. Several of the papers introduced new ideas. J. Connes (in Jacquinot's group) addressed the theoretical line shape and the computational issues of FTS. Strong and his collaborators Vanasse and Gebbie studied the far infrared. Gebbie used an interferometer to measure the solar spectrum between 300 and 1000 cm^{-1} with a resolution of 0.2 cm^{-1} and was able to precisely locate the atmospheric transmission windows. Fellgett obtained the first spectrum of a star in the near-infrared at a resolution between 60 and 100. Mertz constructed a interferometer for the visible region with which stellar spectra were obtained with a resolution of 100.

Connes, J., Domaine d'utilisation de la méthode par transformée de Fourier, *J. Physique et le Radium*, **19**, 197–208, 1958. A detailed discussion of the shape of the instrumental profile, including the effects of finite path length L and of the finite field of view (typically a circular, centered diaphragm).

Gebbie, H. A., *J. Phys. Radium*, **19**, 230, 1958. The first suggestion of numerical apodization in the computation process.

Fellgett, P., *J. Phys. Radium*, **19**, 187, 1958. Initial proposal for the use of cube corners in interferometers. The properties of the cube corner had been discussed previously (Peck, E. R., *J. Opt. Soc. Amer.*, **38**, 1015, 1948, and Murty, M. U. R. K., *J. Opt. Soc. Amer.*, **50**, 7, 1960.

Mertz, L. W., *J. Phys. Radium*, **19**, 233, 1958. A polarization interferometer for the visible spectral region consisting of a double refracting blade of variable thickness between the polarizers.

J. Connes and H. Gush obtained the first *good* high-resolution emission spectrum of the night sky ($R = 1000$) around 6250 cm^{-1} (J. Connes and H. Gush, 1959a; J. Connes and H. Gush, 1959b, 1960; J. Connes, 1960, 1961).

Cat's-eyes as retroreflectors and the stepping method of increasing the path difference were simultaneously introduced by J. and P. Connes (1966). In addition, they demonstrated the multiplex advantage in astronomy with high-spectral-resolution observations of Venus and Mars ($\delta\sigma = 1$ cm^{-1}, $R \sim 1000$) and laboratory spectra with 0.1 cm^{-1} spectral resolution ($R \sim 10\,000$) that are considerably higher quality than state-of-the-art grating spectra at that time. The interferometer was built at the Jet Propulsion Laboratory during a sabbatical visit, tested at Mount Wilson and Steward Observatories, and then taken to the Observatoire de Saint Michel, France.

The fast Fourier transform was discovered, reducing the computational burden for calculating a Fourier transform from an N^2 problem to one of order $N\ln(N)$ (Cooley and Tukey, 1965).

Connes, J., and Gush, H., *Symposium on Interferometry*, Teddington, Great Britain, 1959a.

Connes, J., and Gush, H. P., *J. Phys. Radium*, **20**, 915, 1959b. Balloon-borne FTS for observations of the OH emission in the night sky.

Connes, J., and Gush, H. P., *J. Phys. Radium*, **21**, 615, 1960.

Connes, J., Ph.D. Thesis, Recherches sur la spectroscopie par transfomation de Fourier, Laboratory Aime Cotton, Bellevue, France, 1960.

Connes, J., Recherches sur la spectroscopie par transfomation de Fourier, *Revue d'Optique*, **40**, 45–78, 116–140, 171–190, 231–265, 1961. Published version of J. Connes' thesis presenting the properties of FTS: (a) the theoretical instrumental line shape profile, (b) real instrumental line shape functions, (c) noise in FTS, (d) noise in numerical Fourier transformations, (e) experimental results (night sky spectra). Translated into English by the U.S. Navy (NAV-WEPS Rept. No. 8099, NOTS TP 3157, published by the U.S. Naval Ordnance Test Station, China Lake, CA).

Cooley J. W., and Tukey, J. W., An algorithm for the machine calculation of complex Fourier series, *Mathematics Computation*, 19, 297–301, 1965. The time to transform 10^6 points went down from 60 years to one day (i.e., from totally infeasible to excitingly accessible).

Connes, J., and Connes, P., Near-infrared planetary spectra by Fourier spectroscopy. I. Instruments and results, *J. Opt. Soc. Amer.*, **56**, 896–910, 1966.

Following the Bellevue meeting, the second important meeting "Methodes nouvelles de Spectroscopie Instrumentale" was held in Orsay, France, in April 1966. At this meeting, 20 papers out of 61 were presented on FTS. These papers were published in *Supplément au Journal de Physique (Colloque C2)*, nos. 3–4, **28**, 1967. Forman (1967) introduced the FTS community to the fast Fourier transform (FFT), which arrived none too soon: J. Connes and P. Connes had demonstrated planetary spectra at 0.08 cm^{-1} spectral resolution ($R > 10\,000$), and Pinard demonstrated a spectral resolution of 0.006 cm^{-1} ($R \geq 800\,000$) with a one-meter FTS. Mertz (1967) introduced the notion of phase error and phase correction, and Cuisenier and Pinard (1967) discussed the role of the cat's-eye retroreflector in high-resolution interferometry.

Forman, M. L., Fast Fourier transform technique and its application to Fourier spectroscopy, *J. Opt. Soc. Amer.*, **56**, 978–979, 1966.

Forman, M. L., *J. Physique*, **28**, C2-58, 1967. Introduction of the FFT algorithm to the FTS community.

Mertz, L. W., *J. Physique*, **28**, C2-11, 1967. Initial discussion of phase error.

Cuisenier, M., and Pinard, J., *J. Physique*, **28**, C2-97, 1967. A discussion of the properties of cat's-eye retroreflectors for interferometry.

Connes, P., Connes J., and Maillard, J.-P., Spectroscopie astronomique par transformation de Fourier, *J. Physique*, **28**, C2-120, 1967. A spectrum of Venus at a resolution of 0.08 cm^{-1}.

Pinard, J., *J. Physique*, **28**, C2-136, 1967. The first very high resolution spectra (0.006 cm^{-1}).

The planetary results were consolidated into a reference volume (P. Connes, J. Connes, and J-P. Maillard, 1967).

Rapid scanning and phase correction were more fully developed, and Hanel and coworkers began the development and application of FTS instruments to satellite remote sensing (Hanel and Chaney, 1966, Chaney, *et al.*, 1967).

Phase correction, as a necessary step to obtaining the best possible spectra, remains a central issue in FTS. In the late 1960s two approaches were developed: the Forman method of symmetrizing the interferogram by convolution prior to FFT, and the Mertz method of rotating the complex spectrum into a real spectrum with the phase rotation function. Many years and much paper has been expended arguing the merits of , similarities of, and differences between these methods. The bottom line is that they are mathematically nearly equivalent, but in implementation the Forman method requires one additional apodization function to keep the convolution kernel finite and continuous.

Mertz, L. W., in *Infrared Spectra of Astronomical Bodies* (Proc. 12th Int. Astrophys. Symp., Liège, Belgium, June 1963 (Institute d'Astrophysique, Cointe-Sclessin, Belgium, 1964), p. 120, 1963. Rapid scanning.

Forman, M. L., Steel, W. H., and Vanasse, G. A., Correction of asymmetric interferograms obtained in Fourier spectroscopy, *J. Opt. Soc. Amer.*, **56**, 59–63 1966.

Hanel, R. A., and Chaney, L., Goddard SFC, Rep.X-620-66-476, 1966. NIMBUS III IRIS, with a spectral resolution of 5 cm^{-1}, covering 400 to 2000 cm^{-1}.

Chaney, L., Drayson, S., and Young, C., Fourier transform spectrometer — radiative measurements and temperature inversion, *Appl. Opt.*, **6**, 347, 1967.

Connes, J., Connes, P., and Maillard, J.-P., *Atlas des Spectres Planétaires Infrarouges*, Editions du CNRS, Paris, 1967.

Mertz, L. W., *Infrared Physics*, **7**, 17, 1967. Phase correction.

The Modern Era: 1970-2000

The 1970 Aspen International Conference on Fourier Spectroscopy, March 16-20, 1970, was an important milestone for FTS. It was the first conference entirely devoted to FTS. The proceedings were published *in extenso* as "Special Report No. 114" of the Air Force Cambridge Research Laboratories and are well known within the community. Besides invited papers, several papers represent important landmarks in the development of the field of FTS. The first 1-million point interferograms were presented (Delouis, 1971, and Guelachvili and Maillard, 1971). J. Connes discussed the computational aspects of FTS in detail and gives computational times for various computers of the 1960s. Brault attended the Aspen meeting and decided that FTS was the way to advance laboratory and solar spectroscopy. The notion of a rapid scanning 1-m FTS suitable for both infrared and visible/ultraviolet spectroscopy — the instrument that would become the 1-m FTS at the McMath–Pierce Solar Observatory — was conceived in Aspen.

Delouis H., Fourier transformation of a 10^6 sample interferogram, Aspen International Conference on Fourier Spectroscopy, G. A. Vanasse, A. T. Stair, and D. J. Baker eds., U.S. Air Force Cambridge Research Laboratory, 71-0019, 145–150, 1971.

Guelachvili, G., and Maillard, J.-P., Fourier spectroscopy from 10^6 samples, Aspen International Conference on Fourier Spectroscopy, G. A. Vanasse, A. T. Stair, and D. J. Baker eds., U.S. Air Force Cambridge Research Laboratory, 71-0019, 151–161, 1971.

Connes, J., Computing problems in Fourier spectroscopy, Aspen International Conference on Fourier Spectroscopy, G. A. Vanasse, A. T. Stair, and D. J. Baker eds., U.S. Air Force Cambridge Research Laboratory, 71–0019, 83–115, 1971.

The field of FTS grew rapidly with the following events as significant milestones. In 1970 the first general-purpose commercial instrument (DIGILAB FTS-14s) was introduced, with some 300 units in use over the next seven years. The DIGILAB instrument was preceded by several special-purpose instruments (Grubb Parsons and RIIC in England, and Block Engineering and Idealab in the United States). The growth and application of low-resolution FTS in analytic chemistry has been explosive and has spawned many new techniques. Griffiths and de Haseth provide an introduction and overview of FTS (or FTIR in some communities) and its low-resolution applications in analytic chemistry in the 400- to 4000-cm^{-1} region. The high resolution instruments for laboratory and astronomical spectroscopy continued to improve. For example, spectra of Venus were recorded at 0.015 cm^{-1} resolution. (Connes, 1973). On the long-wavelength side, FTS measurements in the submillimeter and microwave enabled determination of the cosmic background radiation (Richards, 1964, and Woody *et al.*, 1975).

The completion of the 1-m FTS at the National Solar Observatory (now the McMath–Pierce Solar Observatory) in 1976 took the preceding ideas and ambitions (in particular the potential for precision spectrometry in the visible and ultraviolet) and captured them in glass and metal. A bibliography of the spectroscopy completed at NSO has defined the standard for laboratory spectroscopy for the past 25 years. Many preeminent spectroscopy groups mothballed their instruments and began pilgrimages to Kitt Peak.

In terms of frequency and resolution metrics: in 1976, uv spectra were obtained up to 50 000 cm^{-1} by Brault; in 1977, 0.001-cm^{-1} spectral resolution was achieved by Guelachvili; in 1987 FTS entered the vacuum ultra-violet (Irfan *et al.* 1979); in 1985, the first high-resolution FTS reached orbit with the flight of the Atmospheric Trace Molecule Spectroscopy Experiment (ATMOS) on the Space Shuttle (Farmer *et al.*, 1987).

Woody, D. P., Mather, J. C., Nishioka, N. S., and Richards, P. L., Measurement of the spectrum of the submillimeter cosmic background, *Phys. Rev. Lett.*, **34**, 1036–1039, 1975.

Brault, J. W., Solar Fourier transform spectroscopy, *Osservazioni e Memorie dell' Osservatorio Astrofisico di Arcetri* **106**, 33–50, in Proceedings of the JOSO Workshop, G. Godoli, G. Noci, and A. Righini eds., November 7–10, 1978.

Guelachvili, G., High–accuracy Doppler–limited 10^6 samples Fourier transform spectroscopy, *Appl. Opt.*, **17**, 1322–1326, 1978. 0.001-cm^{-1} resolution spectra.

Irfan, A. Y., Thorne, A. P., Bohlander, R. A., and Gebbie, H. A., Fourier transform spectroscopy in the vacuum ultraviolet, *J. Physics E (Scientific Instruments)*, **12**, 6, 472–473, 1979.

Griffiths, P. R., and de Haseth, J. A., in *Fourier Transform Infrared Spectrometry*, P. J. Elving and J. D. Winefordner, eds., Wiley-Interscience, New York, 1986.

Thorne, A. P., Harris, C. J., Wynne-Jones, I., Learner, R. C. M., and Cox, G., A Fourier transform spectrometer for the vacuum ultraviolet: design and performance, *J. Physics E (Scientific Instruments)*, **20**, 1, 54–60, 1987.

Farmer, C. B., High-resolution infrared spectroscopy of the Sun and the Earth's atmosphere from space, *Mikrochim. Acta [Wien]*, **III**, 189–214, 1987. [ATMOS]

Farmer, C. B., Raper, O. F., and O'Callaghan, F. G., Final report on the first flight of the ATMOS instrument during the Spacelab 3 mission, April 29 through May 6, 1985, 1987, Jet Propulsion Laboratory Publication 87 32, 1987.

Farmer, C. B., The ATMOS solar atlas, *Infrared Solar Physics*, 511–521, 1994.

Brault, J. W., New approach to high-precision Fourier transform spectrometer design, *Appl. Opt.*, **35**, 16, 2891–2896, 1996.

14

APPLICATIONS BIBLIOGRAPHY

High-Resolution Laboratory Measurements

Numerous compendia of line parameters have been generated over the last 40 years, and they continue to be refined periodically. For atmospheric transmission calculations these include the U.S. Air Force Geophysical Laboratory (AFGL) and HITRAN (high-resolution molecular absorption) line parameter databases (Rothman *et al.* 1983a, 1983b, 1987, 1992, and 1998), the Atmospheric Trace Molecule Spectroscopy (ATMOS) experiment linelist (Brown *et al.*, 1987, and Brown *et al.*, 1996); the GEISA (Gestion et Etude des Informations Spectroscopiques Atmospheriques) program (Husson *et al.*, 1992). It is notable that FTS laboratory measurements predominate within these databases.

Primary and secondary calibration standards in the infrared, visible, and ultraviolet continue to be updated, corresponding to the state of the art in spectrometry (Brown *et al., 1983, Johns 1987; Learner and Thorne 1988; Nave et al.*, 1991, 1992; Whaling *et al.*, 1995; Guelachvili *et al.*, 1996).

Brown, L. R., Margolis, J. S., Norton, R. H., and Stedry, B. A., Computer measurements of line strengths with application to the methane spectrum, *Appl. Spectrosc.*, **37**, 287–292, 1983.

Brown, L. R., Farmer, C. B., Rinsland, C. P., and Toth, R. A., Molecular line parameters of the atmospheric trace molecule spectroscopy experiment, *Appl. Opt.*, **26**, 5154–5182, 1987.

Brown, L. R., Gunson, M. R., Toth, R. A., Irion, F. W., Rinsland, C. P., and Goldman, A., The 1995 Atmospheric Trace Molecule Spectroscopy (ATMOS) linelist, *Appl. Opt.*, **35**, 2828–2848, 1996.

Guelachvili, G., Birk, M., Borde, C. J., Brault, J. W., Brown, L. R., Carli, B., Cole, A. R. H., Evenson, K. M., Fayt, A., Hausamann, D., Johns, J. W. C., Kauppinen, J., Kou, Q., Maki, A. G., Narahari Rao, K., Toth, R. A., Urban, W., Valentin, A., Verges, J., Wagner, G., Wappelhorst, M. H., Wells, J. S., Winnewisser, B. P., and Winnewisser, M., High-resolution wavenumber standards for the infrared, *Spectrochimica Acta, Part A (Molec. and Biomolec. Spectrosc.)*, **52A**, 7, 717–732, 1996.

Husson, N., Bonnet, B., Scott, N. A., and Chedin, A., Management and study of spectroscopic information: the GEISA program, *J. Quant. Spectrosc. Radiat. Transfer*, **48**, 509–518, 1992.

Johns, J. W., High resolution and the accurate measurement of intensities, *Mikrochim. Acta [Wien]*, **III**, 171–188, 1987.

Learner, R. C. M., and Thorne, A. P., Wavelength calibration of Fourier transform emission spectra with applications to Fe I, *J. Opt. Soc. Amer. B*, **5**, 2045–2049, 1988.

Nave, G., Learner, R. C. M., Thorne, A. P., and Harris, C. J., Precision Fe I and Fe II wavelengths in the ultraviolet spectrum of the iron-neon hollow cathode lamp, *J. Opt. Soc. Amer. B*, **8**, 2028–2041, 1991.

Nave, G., Learner, R. C. M., Murray, J. E., Thorne, A. P. and Brault, J. W., Precision Fe I and Fe II wavelengths in the red and infrared spectrum of the iron-neon hollow cathode lamp, *J. Phys. II (France)*, **2**, 913–929, 1992.

Rothman, L. S., Goldman, A., Gillis, J. R., Gamache, R. R., Pickett, H. M., Poynter, R. L., Husson, N., and Chedin, A., AFGL trace gas compilation — 1982 version, *Appl. Opt.*, **22**, 1616–1627, 1983a.

Rothman, L. S., Gamache, R. R., Barbe, A., Goldman, A., Gillis, J. R., Brown, L. R., Toth, R. A., Flaud, J.-M., and Camy-Peyret, C., AFGL atmospheric absorption-line parameters compilation — 1982 edition, *Appl. Opt.*, **22**, 2247–2256, 1983b.

Rothman, L. S., Gamache, R. R., Goldman, A., Brown, L. R., Toth, R. A., Pickett, H. M., Poynter, R. L., Flaud, J.-M., Camy-Peyret, C., Barbe, A., Husson, N., Rinsland, C. P., and Smith, M. A. H., The HITRAN database: 1986 edition, *Appl. Opt.*, **26**, 4058–4097, 1987.

Rothman, L. S., Gamache, R. R., Tipping, R. H., Rinsland, C. P., Smith, M. A. H., Benner, D. C., Malathy Devi, V., Flaud, J.-M., Camy-Peyret, C., Perrin, A.,

Goldman, A., Massie, S. T., Brown, L. R., and Toth, R. A., The HITRAN molecular database editions of 1991 and 1992, *Appl. Opt.*, **26**, 4058–4097, 1992.

Rothman, L. S., Rinsland, C. P., Goldman, A., Massie, S. T., Flaud, J.-M., Perrin, A., Dana, V., Mandin, J.-Y., Schroeder, J., Mc Cann, A., Gamache, R. R., Wattson, R. B., Yoshino, K., Chance, K., Jucks, K., Brown, L. R. and Varanasi, P., The HITRAN molecular spectroscopic database and HAWKS (HITRAN Atmospheric Workstation), *J. Quant. Spectrosc. Radiat. Transfer*, **60**, 665–682, 1998.

Whaling, W., Anderson, W. H. C., Carle, M. T., Brault, J. W., Zarem, H. A., Argon ion linelist and level energies in the hollow-cathode discharge, *J. Quant. Spectrosc. Radiat. Transfer*, **53**, 1, 1–22, 1995.

Solar and Stellar Astronomy

Brault, J. W., A high-precision Fourier spectrometer for the visible, *Auxiliary Instrumentation for Large Telescopes*, 367–373, CERN Conference Proceedings, S. Laustsen and A. Reiz eds., Geneva, Switzerland, May 2–5, 1972.

Brault, J. W., Solar Fourier transform spectroscopy, *Osservazioni e Memorie dell' Osservatorio Astrofisico di Arcetri*, **106**, 33–50, 1978, G. Godoli, G. Noci, and A. Righini eds., Proceedings of the JOSO Workshop, November 7–10, 1978.

Delbouille, L., Roland, G., Brault, J. W., and Testerman, L., *Photometric Atlas of the Solar Spectrum from 1,850 to 10,000* cm^{-1}, Kitt Peak National Observatory Publication, National Optical Astronomy Observatory, 1981.

Delbouille, L., Roland, G., and Gebbie, H. A., Actual possibilities of astronomical infrared spectroscopy by interferometry and Fourier transformation, *Ap. J.*, **69**, 334–338, 1964.

Farmer, C. B., Delbouille, L., Roland, G., and Servais, C., *The Solar Spectrum Between 16 and 40 microns*, San Juan Capistrano Research Inst. Publication; SJI Tech. Report 94-2, 1994.

Hall, D. N. B., Ridgway, S. T., Bell, E. A. and Yarborough, J. M., A 1.4 Meter Fourier Transform Spectrometer for astonomical observations, *Proc. SPIE*, **172**,121–129, 1979.

Jennings, D. E., Hubbard, R., and Brault, J. W., Double passing the Kitt Peak 1-Meter Fourier transform spectrometer, *Appl. Opt.*, **24**, 3438–3440, 1985.

Maillard, J.-P., and Michel, G., A high-resolution Fourier transform spectrometer for the cassegrain focus at the CFH telescope, in *Instrumentation for Astronomy with Large Optical Telescopes*, C. M. Humphries ed., 213–222, Kluwer Academic Reidel, Norwell, MA, 1982.

Ridgway, S. T., and Brault, J. W., Astronomical Fourier transform spectroscopy revisited, *Ann. Rev. of Astron. Astrophys.*, **22**, 291–317, 1984.

Atmospheric Remote Sensing

Beer, R., Norton, R. H., and Seeman, C. H., *Rev. Sci. Instrum.*, **42**, 1393–1403, 1971.

Carli, B., Mencaraglia, F., and Bonetti, A., Submillimeter high- resolution FT spectrometer for atmospheric studies, *Appl. Opt.*, **23**, 2594–2603, 1984.

Connes, J., and Gush, H. P., *J. Phys. Radium*, **20**, 915, 1959. Balloon-borne FTS for observations of the OH emission in the night sky.

Delbouille, L., The role of very high-resolution FTS in recent determinations of the quantitative composition of the stratosphere, *Proc. SPIE*, **2089**, 32–37, 1994.

Farmer, C. B., *Can. J. Chem*, **52**, 1544–1559, 1974.

Gush, H. P., and Buijs, H. L., *Can. J. Phys.*, **42**, 1037, 1964. Balloon-borne FTS for observations of emission by OH in the night sky.

Kunde, V. G., Brasunas, J. C., Conrath, B. J., Hanel, R. A., Herman, J. R., Jennings, D. E., Maguire, W. C., Walser, D. W., Annen, J. N., Silverstein, M. J., Abbas, M. M., Herath, L. W., Buijs, H. L., Berube, J. N., and McKinnon, J., Infrared spectroscopy of the lower atmosphere with a balloon-borne Fourier spectrometer, *Appl. Opt.*, **26**, 545–553, 1987. Mid-infrared limb emission atmospheric sounding with a cryogenic FTS.

Schinder, R. A., *Appl. Opt.*, **9**, 301–306, 1970.

Toth, R. A., Farmer, C. B., Schindler, R. A. Raper, O. F., Shaper, P. W., *Nature (London) Phys. Sci.*, **244**, 7–8, 1973.

Toon, G. C., The JPL MkIV interferometer, *Optics and Photonics News*, **2**, 19-21, October 1991.

Traub, W. A., Chance, K. V., Brasunas, J. C., Vrtilek, J. M., and Carleton, N. P., Use of a Fourier transform spectrometer on a balloon-borne telescope and at the Multiple Mirror Telescope (MMT), *Proc. SPIE*, **331**, 208–218, 1982.

Woody, D. P., Mather, J. C., Nishioka, N. S., and Richards, P. L., Measurement of the spectrum of the submillimeter cosmic background, *Physical Review Letters*, **34**, 1036–1039, 1975.

Space-Based Remote Sensing

The remarkable capability of the FTS for serendipitous discoveries is best illustrated by the FTS instruments developed for planetary exploration by Hanel and coworkers at NASA Goddard Space Flight Center. They were successfully demonstrated on two Earth-observing missions (NIMBUS III/IV, 1969/1970) for atmospheric and surface composition characterization, the Infrared Interferometer Spectrometer (IRIS) instruments were the first of the long-duration, continuously operating FTS systems (Hanel *et al.* 1969, 1970, 1971, 1972a). Subsequent missions to Mars (Mariner 9, 1971; Hanel *et al.*, 1972b, 1972c), Jupiter and Saturn (Voyager I, 1977–1980, Hanel *et al.*, 1979, 1980), and Jupiter, Saturn, Uranus, and Neptune (Voyager II, 1977–1989, Hanel *et al.*, 1986; Courtin *et al.*, 1984, and Coustenis *et al.*, 1989) redefined our understanding of the surface and atmospheric composition of the planets and their satellites.

The discovery of a chemically complex atmosphere around Titan by Voyager was totally unexpected and has led to the design and flight of the Composite Infrared Spectrometer (CIRS) an imaging FTS on CASSINI (launched in 1997) to characterize the surface and atmosphere of Titan (Maymon *et al.*, 1993, and Kunde *et al.*, 1996). The Thermal Emission Spectrometer (TES/M) on the Mars Global Surveyor (1998) has provided excellent surface maps throughout the mission's lifetime (Christensen *et al.*, 1992, 1998, and Bandfield *et al.*, 2000). On each of the planetary missions the FTS systems continued to operate until shutdown at the end of the mission.

Bandfield, J. L., Hamilton, V. E., and Christensen, P. R., A global view of Martian surface compositions from MGS-TES, *Science*, **287**, 1626–1630, 2000.

Christensen, P. R., Anderson, D. L., Stillman, C. C., Clark, R. N., Kieffer, H. H., Malin, M. C., Pearl, J. C., Carpenter, J. C., Bandera, N. B., Brown, F. G., and Silverman, S., Thermal Emission Spectrometer experiment: Mars Observer mission, *J. Geophys. Res.*, **97**, 7719–7734, 1992.

Christensen, P.R., D.L. Anderson, S.C. Chase, R.T. Clancy, R.N. Clark, B.J. Conrath, H.H. Kieffer, R.O. Kuzman, M.C. Malin, J.C. Pearl, T.L. Roush, and M.D. Smith, Initial Results from the Mars Global Surveyor Thermal Emission Spectrometer Investigation, *Science*, **279**, 1692–1698, 1998.

Conrath, B. J., Flasar, F. M., Hanel, R. A., Kunde, V. G., Maguire, W., Pearl, J., Pirraglia, J., Samuelson, P., Gierasch, P., Weir, A., Bézard, B., Gautier, D., Cruikshank, D., Horn, L., Springer, R., Shaffer, W., Infrared observations of the Neptunian system, *Science*, **246**, 1454–1459, 1989.

Courtin, R., Gautier, D., Marten, A., Bézard, B., and Hanel, R. A., The composition of Saturn's atmosphere at northern temperate latitudes from Voyager IRIS spectra: NH_3, PH_3, C_2H_2, C_2H_4, CH_3D, CH_4, and the Saturnian D/H isotopic ratio, *Astrophys. J.*, **287**, 899–916, 1984.

Coustenis, A., Bézard, B, and Gautier, D., Titan's atmosphere from Voyager infrared observations I. The gas composition of Titan's equatorial region, *Icarus*, **80**, 54–76, 1989.

Hanel, R. A., and Conrath, B., Interferometer experiment on Nimbus 3: preliminary results, *Science*, **204**, 972–976, 1969.

Hanel, R. A., Schlachman, B., Clark, F. D., Prokesh, C. H., Taylor, J. B., Wilson, W. M., and Chaney, L., The Nimbus 3 Michelson interferometer, *Appl. Opt.*, **9**, 1767–1774, 1970.

Hanel, R. A., Schlachman, B., Rodgers, D., and Vanous, D., The Nimbus 4 Michelson interferometer, *Appl. Opt.*, **10**, 1376–1381, 1971.

Hanel, R. A., Conrath, B. J., Kunde, V. G., Prabhakara, C., Revah, I., Salomonson, V. V., and Wolford, G., The Nimbus 4 infrared spectroscopy experiment, I. Calibrated thermal emission spectra, *J. Geophys. Res.*, **77**, 2639–2641, 1972a.

Hanel, R. A., Schlachman, B., Breihan, E., Bywaters, R., Chapman, F., Rhodes, M., Rogers, D., and Vanous, D., The Mariner 9 Michelson interferometer, *Appl. Opt.*, **11**, 2625–2634, 1972b.

Hanel, R. A., Conrath, B. J., Hovis, W. A., Kunde, V. G., Lowman, P. D., Pearl, J. C., Prabhakara, C., and Schlachman, B. L., Infrared spectroscopy experiment on the Mariner 9 mission: preliminary results, *Science*, **175**, 305–308, 1972c.

Hanel, R. A., Conrath, B. J., Flasar, M., Kunde, V. G., Lowman, P., Maguire, W., Pearl, J., Pirraglia, J., and Samuelson, R., Infrared observations of the Jovain system from Voyager I, *Science*, **204**, 972–976, 1979.

Hanel, R. A., Crosby, D., Herath, L., Vanous, D., Collins, D., Creswick, H., Harris, C., and Rhodes, M., Infrared spectrometer for Voyager, *Appl. Opt.*, **19**, 1391–1400, 1980.

Hanel, R. A., Conrath, B. J., Flasar, F. M., Kunde, V. G., Maguire, W., Pearl, J., Pirraglia, J., Samuelson, R., Cruikshank, D., Gautier, D., Gierasch, P., Horn, L., and Schulte, P., Infrared observations of the Uranian system, *Science*, **233**, 70–74, 1986.

Kunde, V. G., *et al.*, Cassini infrared Fourier spectroscopic investigation, *Proc. SPIE*, **2803**, Cassini/Huygens: A Mission to the Saturnian Systems, Linda Horn ed., pp. 162–177, 1996.

Maymon, P. W., Dittman, M. G., Pasquale, B. A., Jennings, D. E., Mehalick, K. I., Trout, C. J., Optical design of the Composite Infrared Spectrometer (CIRS) for the CASSINI mission, *Proc. SPIE*, **1945**, Space Astronomical Telescopes and Instruments II, P. Y. Bely, and J. B. Breckinridge eds. pp. 100–111, 1993.

In astronomy, the FTS has been a key enabling technique, in particular for measurements of the cosmic microwave background (CMB). The Far InfraRed Absolute Spectrophotometer (FIRAS) was one of three instruments on the Cosmic Background Explorer (Mather and Kelsall, 1980; Mather *et al.*, 1993, 1994), which compared the CMB radiation to an accurate blackbody and observed the spatial distribution of dust and line emission from the galaxy. For mid-infrared astronomy, the Spatial Infrared Imaging Telescope (SPIRIT III) sensor, developed for the Midcourse Space Experiment (MSX, 1996–99) by the Ballistic Missile Defense Organization (BMDO) demonstrated an FTS system for 2 to 28 micrometer astronomical spectroscopy (Mill, 1994; Zachor *et al.*, 1994). In 1998, ESA selected the Spectral and Photometric Imaging Receiver (SPIRE), which included an IFTS for the FIRST satellite (Griffin, 1998), and Japan is developing the Far-Infrared Surveyor (FIS), also containing an IFTS, for the IR Imaging Surveyor (IRIS) satellite (Kawada, 1998).

Griffin, M. J., Bolometer instrument for FIRST, *Proc. SPIE*, Advanced Technology, MMW, Radio, and Terrahertz Telescopes, T. G. Phillips ed., **3357**, 45, 1998.

Kawada, M., FIS: Far-Infrared Surveyor onboard IRIS, *Proc. SPIE*, Infrared Astronomical Instrumentation, A. M. Fowler ed., **3354**, 6, 1998.

Mather, J., and Kelsall, T., The Cosmic Background Explorer Satellite, *Physica Scripta*, **21**, 671–677, 1980.

Mather, J. C., Fixsen D. J., and Shafer, R. A., Design for the COBE Far-Infrared Absolute Spectrophotometer (FIRAS), *Proc. SPIE*, **2019**, 168, 1993.

Mather, J. C., Cheng, E. S, Cottingham, D. A., Eplee, R. E. Jr., Fixsen, D. J., Hewagama, T., Isaacman, R. B., Jensen, K. A., Meyer, S. S., Noerdlinger, P. D., Read, S. M., Rosen, L. P., Shafer, R. A., Wright, E. L., Bennett, C. L., Boggess, N. W., Hauser, M. G., Kelsall, T., Moseley, S. H. Jr., Silverberg, R. F., Smoot, G. F., Weiss, R., and Wilkins, D. T., Measurement of the cosmic microwave background spectrum by the COBE FIRAS instrument, *Astrophys. J.*, **420**, 439, 1994.

Mill, J. D., Midcourse Space Experiment (MSX), an overview of instruments and data collection plan: signal processing, sensor fusion, and target recognition III, *Proc. SPIE*, **2232**, 200, 1994.

Zachor A. *et al.*, Mid-course Space Experiment (MSX): capabilities of the LWIR interferometer for remote sensing of trace constituents in the stratosphere and mesosphere; Atmospheric Propagation and Remote Sensing, *Proc. SPIE*, **2222**, 99, 1994.

In applications to terrestrial remote sensing, numerous FTS instruments have been developed and operated, yielding a vast amount of information about the state, composition, and dynamics of Earth's atmosphere, ocean, and surface. Nadir sounders, operating in the 3- to 15-micrometer spectral region were demonstrated on the Nimbus program (Nimbus III, 1969, and Nimbus IV, 1970) and have become the mainstay of infrared sensors for meteorological and climate research: the Interferometric Monitor of Greenhouse Gases (IMG, Shimoda and Ogawa, 1994) on the Japanese ADEOS satellite (1996–97), the Michelson Interferometer for Passive Atmospheric Sounding (MIPAS, de Zoeten *et al.*, 1993; Endemann *et al.*, 1994), developed for the European ENVISAT satellite, the Tropospheric Emission Sounder (TES, Beer and Glavitch, 1989), under development for the Earth Observing System (EOS), the Interferometric Atmospheric Sounding Interferometer (IASI; Javelle and Cayla 1994), under development for the European METOP operational satellite program, and the Cross-track Infrared Sounder (CrIS) for the U.S. National Polar Orbiting Environmental Satellite System (NPOESS).

Beer, R., and Glavitch, T. A., Remote sensing of the troposphere by infrared emission spectroscopy; Advanced Optical Instrumentation for Remote Sensing of the Earth's Surface from Space, *Proc. SPIE*, **1129**, 42-51, 1989. [TES/EOS CHEM]

de Zoeten, P., Haurer, R., and Birkl, R., Optical design of the Michelson Interferometer for Passive Atmospheric Sounding; Passive Infrared Remote Sensing of Clouds and the Atmosphere, *Proc. SPIE*, **1934**, 284, 1993.

Endemann, M., Lange, G., and Fladt, B., Michelson Interferometer for Passive Atmospheric Sounding; Space Optics 1994: Earth Observation and Astronomy. *Proc. SPIE*, **2209**, 36, 1994. [MIPAS/ENVISAT]

Javelle, P., and Cayla, F., Infrared Atmospheric Sounding Interferometer (IASI) instrument overview; Space Optics 1994: Earth Observation and Astronomy, *Proc. SPIE*, **2209**, 14, 1994. [IASI/METOP]

Shimoda, H., and Ogawa, T., Interferometric Monitor for Greenhouse Gases (IMG); Infrared Spaceborne Remote Sensing II, *Proc. SPIE*, **2268**, 92, 1994.

In the period between 1979 and 1995, the Atmospheric Trace Molecule Spectroscopy (ATMOS) Experiment, an FTS instrument was developed for solar occultation limb sounding on the Space Shuttle (Farmer, 1987; Farmer *et al.*, 1987; Farmer and Norton, 1989a and b; Farmer, 1994; and Abrams *et al.*, 1996) and will be continued by the Canadian SCISAT-1 Atmospheric Chemistry Experiment (SCISAT-1/ACE, scheduled for launch in 2002). Key to the deployment of FTS systems for space was the development of "aligned-by-design" optical systems that relax alignment tolerances and stability from arc-seconds to arc-minutes, which renders the instruments insensitive to mechanical or thermally induced misalignment (Farmer, 1987). The ATMOS interferometer flew on the Space Shuttle four times (1985, 1992, 1993, and 1994) and did not require realignment, due to its optical configuration, between fabrication in 1982 and decommissioning in 1999.

ATMOS was state of the art when it was built in 1982, and only in 1998 were we able to demonstrate sufficient improvement in designs to warrant moving to a free-flying version of this sensor. An indication of the technology readiness of FTS systems for spaceflight applications is the mass and volume reductions possible between ATMOS and ACE: The mass decreased from 250 kg to 29 kg and the volume decreased from 1 m^3 to less than 0.02 m^3 without compromising the scientific capability of the sensor.

Abrams, M. C., Gunson, M. R., Chang, A. Y., Rinsland, C. P., and Zander, R., Remote sensing of the Earth's atmosphere from space with high-resolution Fourier-transform spectroscopy: development and methodology of data processing for the Atmospheric Trace Molecule Spectroscopy experiment, *Appl. Opt.*, **35**, 2776–2786, 1996.

Farmer, C. B., High-resolution infrared spectroscopy of the Sun and the Earth's atmosphere from space, *Mikrochim. Acta [Wien]*, **III**, 189–214, 1987.

Farmer, C. B., The ATMOS solar atlas, *Infrared Solar Physics*, 511–521, 1994.

Farmer, C. B., Raper, O. F., and O'Callaghan, F. G., Final report on the first flight of the ATMOS instrument during the Spacelab 3 mission, April 29 through May 6, 1985, Jet Propulsion Laboratory Publication 87 32, 1987.

Farmer, C. B., and Norton, R. H., Atlas of the infrared spectrum of the Sun and the Earth atmosphere from space. Volume I, The Sun, NASA Reference Publication 1224, 1989a.

Farmer, C. B., and Norton, R. H., Atlas of the infrared spectrum of the Sun and the Earth atmosphere from space. Volume II, Stratosphere and Mesosphere, 650 to 3350 cm^{-1}, NASA Reference Publication 1224, 1989b.

Geller, M., Atlas of the infrared spectrum of the Sun and the Earth atmosphere from space. Volume III, Key to Identification of Solar Features, NASA Reference Publication 1224, 1989.

15

AUTHOR BIBLIOGRAPHY

Abrams, M. C., High-precision Fourier transform spectroscopy of atmospheric molecules, Ph.D. thesis, University of California at Berkeley, Berkeley, CA, April 1990. A stepping stone between the "Saas Fee Notes" and this book, with a detailed discussion of resolution enhancement of FTS spectra correcting for distortion of the spectrum by the instrument line shape function.

Abrams, M. C., Toon, G. C., and Schindler, R. A., Practical example of the correction of Fourier-transform spectra for detector nonlinearity, *Appl. Opt.*, **33**, 27, 6307–6314, 1994. N. B.: Subsequent to publication we determined that the *nonlinearity* attributed to the MCT detector was actually introduced by the amplifiers. Removal of the amplifiers essentially eliminated the nonlinearity and did not decrease the signal-to-noise ratio (MCA).

Abrams, M. C., Gunson, M. R., Chang, A. Y., Rinsland, C. P., and Zander, R., Remote sensing of the Earth's atmosphere from space with high-resolution Fourier-transform spectroscopy: development and methodology of data processing for the Atmospheric Trace Molecule Spectroscopy experiment, *Appl. Opt.*, **35**, 2776–2786, 1996.

Bandfield, J. L., Hamilton, V. E., and Christensen, P. R., A global view of Martian surface compositions from MGS-TES, *Science*, **287**, 1626–1630, 2000.

Bartlett, M. S., Periodogram Analysis and Continuous Spectra, *Biometrika*, **37**, 1–16, 1950.

Beer, R., Norton, R. H., and Seeman, C. H., *Rev. Sci. Instrum.*, **42**, 1393–1403, 1971.

Beer, R., *Remote Sensing by Fourier Transform Spectrometry*, Wiley, New York, 1992.

Beer, R., and Cayford, A. H., An investigation of a fundamental intensity error in Fourier spectroscopy, *J. Physique*, **28**, C2-38–39, 1967.

Beer, R., and Glavitch, T. A., Remote sensing of the troposphere by infrared emission spectroscopy, Advanced Optical Instrumentation for Remote Sensing of the Earth's Surface from Space, *Proc. SPIE*, **1129**, 42–51, 1989.

Bennett, C. L., Fourier transform IR measurements of thermal infrared sky radiance and Transmission, *Proc. SPIE*, 2266, 25–35, 1994.

Bennett, C. L., Effect of jitter on an imaging FTIR spectrometer, *Proc. SPIE*, 3063, 174–184, 1997.

Bennett, C. L., Carter, M. R., Fields, D. J., Hyperspectral imaging in the infrared using LIFTIRS, *Proc. SPIE*, 2552, 274–283, 1995.

Bennett, C. L., Carter, M. R., Fields, D. J., Lee, F. D., Infrared hyperspectral imaging results from vapor plume experiments, *Proc. SPIE*, 2480, 435–444, 1995.

Bennett, C. L., Carter, M. R., Fields, D. J., Hernandez, J. A., Imaging Fourier transform spectrometer, *Proc. SPIE*, 1937, 191–200, 1993.

Birk, M., and Brault, J. W., Detector quantum efficiency: an important parameter for FT-IR spectroscopy, *Mikrochim. Acta [Wien]*, **II**, 243–247, 1988.

Blackman R. B., and Tukey, J. W., *The Measurement of Power Spectra*, Dover, New York, 1958.

Bracewell, R. N., *The Fourier Transform and Its Applications*, McGraw Hill, New York, 1965.

Bracewell, R. N., Numerical transforms, *Science*, **248**, 647–704, 1990.

Brault, J. W., A High-precision Fourier spectrometer for the visible, *Auxiliary Instrumentation for Large Telescopes*, pp. 367–373, CERN Conference Proceedings, S. Laustsen and A. Reiz eds., Geneva, Switzerland, May 2–5, 1972.

Brault, J. W., Solar Fourier transform spectroscopy, *Osservazioni e Memorie dell' Osservatorio Astrofisico di Arcetri* **106**, 33–50, G. Godoli, G. Noci, and A. Righini eds., Proceedings of the JOSO Workshop, November 7–10, 1978.

Brault, J. W., Fourier transform spectrometry in relation to other passive spectrometers, *Phil. Trans. R. Soc. Lond*, A, **307**, 503, 1982.

Brault, J. W., Fourier transform spectrometry, in *High Resolution in Astronomy*, Proceedings of the Fifteenth Advanced Course of the Swiss Society of Astronomy and Astrophysics, A. O. Benz, M. C. E. Huber, and M. Mayor, eds., Saas Fee, 1985 (Sauverny, Observatoire de Genève, Switzerland), pp. 1–61. Often referred to as the "Saas Fee" notes. They are the seminal publication on which this book is based.

Brault, J. W., High-precision Fourier transform spectrometry: the critical role of phase corrections, *Mikrochim. Acta [Wien]*, **III**, 215, 1987.

Brault, J. W., New approach to high-precision Fourier transform spectrometer design, *Appl. Opt.*, **35**, 16, 2891–2896, 1996.

Brault J. W., and White, O. R., The analysis and restoration of astronomical data via the fast Fourier transform, *Astron. and Astrophys.*, **13**, 169–189, 1971. A thorough introduction to optimum filtering for the restoration of spectra using the FFT, as a historical note, this paper precedes Brault's work in FTS and is focused on the restoration of grating spectra.

Brown, L. R., Margolis, J. S., Norton, R. H., and Stedry, B. A., Computer measurements of line strengths with application to the methane spectrum, *Appl. Spectrosc.*, **37**, 287–292, 1983.

Brown, L. R., Farmer, C. B., Rinsland, C. P., and Toth, R. A., Molecular line parameters of the atmospheric trace molecule spectroscopy experiment, *Appl. Opt.*, **26**, 5154–5182, 1987.

Brown, L. R., Gunson, M. R., Toth, R. A., Irion, F. W., Rinsland, C. P., and Goldman, A., The 1995 Atmospheric Trace Molecule Spectroscopy (ATMOS) linelist, *Appl. Opt.*, **35**, 2845, 1996.

Carli, B., Mencaraglia, F., and Bonetti, A., Submillimeter high- resolution FT spectrometer for atmospheric studies, *Appl. Opt.*, **23**, 2594–2603, 1984.

Carter, M. R., Bennett, C. L., Fields, D. J., Hernandez, J. A., Gaseous effluent monitoring and identification using an imaging Fourier transform spectrometer, *Proc. SPIE*, 2092, 16–26, 1994.

Chamberlain, J., *The Principles of Interferometric Spectroscopy*, Wiley, New York, 1979. Fourier transform spectrometry from a spectroscopist's point of view.

Chaney, L., Drayson, S., and Young, C., Fourier transform spectrometer — radiative measurements and temperature inversion, *Appl. Opt.*, **6**, 347, 1967.

Christensen, P. R., Anderson, D. L., Stillman, C. C., Clark, R. N., Kieffer, H. H., Malin, M. C., Pearl, J. C., Carpenter, J. C., Bandera, N. B., Brown, F. G., and Silverman, S., Thermal Emission Spectrometer experiment: Mars Observer mission, *J. Geophys. Res.*, **97**, 7719–7734, 1992.

Christensen, P.R., D.L. Anderson, S.C. Chase, R.T. Clancy, R.N. Clark, B.J. Conrath, H.H. Kieffer, R.O. Kuzman, M.C. Malin, J.C. Pearl, T.L. Roush, and M.D. Smith, Initial Results from the Mars Global Surveyor Thermal Emission Spectrometer Investigation, *Science*, **279**, 1692–1698, 1998.

Cochran, W. T., Cooley, J. W., Favin. D. L., Helms, H. D., Kaenel, R. A., Lang., W. W., Maling, G. C., Nelson., D. E., Rader, C. M., Welch, P. D., *IEEE Trans. Audio Electroacoustics, Special Issue on Fast Fourier Transform and Its Application to Digital Filtering and Spectral Analysis*, AU-15, **2**, 45, 1967. A complete discussion of FFT algorithms and applications — the definitive reference.

Connes, J., Domaine d'utilisation de la méthode par transformée de Fourier, *J. Physique et le Radium*, **19**, 197–208, 1958. A detailed discussion of the shape of the instrumental profile including the effects of finite path length L and of the finite field of view (typically a circular, centered diaphragm, optimum diaphragm).

Connes, J., *Recherches sur la spectroscopie par transfomation de Fourier*, Ph.D. Thesis, Laboratory Aime Cotton, Bellevue, France, 1960.

Connes, J., Recherches sur la spectroscopie par transfomation de Fourier, *Revue d'Optique*, **40**, 45–78, 116–140, 171–190, 231–265, 1961. Published version of J. Connes' thesis presenting the properties of FTS: (a) the theoretical instrumental line shape profile, (b) real instrumental line shape functions, (c) noise in FTS, (d) noise in numerical Fourier transformations, (e) experimental results (night sky spectra). Translated into english by the U.S. Navy (NAV-WEPS Rept. No. 8099, NOTS TP 3157, published by the U.S. Naval Ordnance Test Station, China Lake, CA).

Connes, J., Computing problems in Fourier spectroscopy, *Aspen International Conference on Fourier Spectroscopy*, G. A. Vanasse, A. T. Stair, and D. J. Baker eds., 139–142, U.S. Air Force Cambridge Research Laboratory, 71-0019, 83-115, 1971.

Connes, J., and Connes, P., Near-infrared planetary spectra by Fourier spectroscopy. I. Instruments and results, *J. Opt. Soc. Amer.*, **56**, 896–910, 1966.

Connes, J., Connes, P., and Maillard, J.-P., *Atlas des Spectres Planétaires Infrarouges*, Editions du CNRS, Paris, 1967.

Connes, J., and Gush, H., *Symposium on Interferometry*, Teddington, Great Britain, 1959a.

Connes, J., and Gush, H. P., *J. Phys. Radium*, **20**, 915, 1959b. Balloon-borne FTS for observations of the OH emission in the night sky.

Connes, J., and Gush, H. P., *J. Phys. Radium*, **21**, 615, 1960.

Connes, P., Connes J., and Maillard, J.-P., Spectroscopie astronomique par transformation de Fourier, *Journal de Physique*, **28**, C2-120, 1967. A spectrum of Venus at a resolution of 0.08 cm^{-1}.

Connes, P., Astronomical Fourier spectroscopy, *Ann. Rev. Astron. Astrophys.*, **8**, 209, 1970. A good early review of the role of Fourier spectrometry in astronomy.

Conrath, B. J., Flasar, F. M., Hanel, R. A., Kunde, V. G., Maguire, W., Pearl, J., Pirraglia, J., Samuelson, P., Gierasch, P., Weir, A., Bézard, B., Gautier, D., Cruikshank, D., Horn, L., Springer, R., Shaffer, W., Infrared observations of the Neptunian system, *Science*, **246**, 1454–1459, 1989.

Cooley J. W., and Tukey, J. W., An algorithm for the machine calculation of complex Fourier series, *Mathematics Computation*, **19**, 297–301, 1965. The time to transform 10^6 points went down from one day to one second.

Courtin, R., Gautier, D., Marten, A., Bézard, B., and Hanel, R. A., The composition of Saturn's atmosphere at northern temperate latitudes from Voyager IRIS spectra: NH_3, PH_3, C_2H_2, C_2H_4, CH_3D, CH_4, and the Saturnian D/H isotopic ratio, *Astrophys. J.*, **287**, 899–916, 1984.

Coustenis, A., Bézard, B, and Gautier, D., Titan's atmosphere from Voyager infrared observations I. The gas composition of Titan's equatorial region, *Icarus*, **80**, 54–76, 1989.

Cuisenier, M., and Pinard, J., *Journal de Physique*, **28**, C2-97, 1967. A discussion of the properties of cat's-eye retroreflectors for interferometry.

Davis, S. P., *Diffraction Grating Spectrographs*, Holt, Rinehart, and Winston, New York, 1970.

Davis, S. P., Laboratory spectroscopy of astrophysically interesting molecules, *Proc. Astron. Soc. Pac.*, **99**, 1105, 1987.

Davis, S. P., and Littleton, J. E., Line intensity determinations for the Red System of CaH, unpublished, 1989.

Delbouille, L., The role of very high resolution FTS in recent determinations of the quantitative composition of the stratosphere, *Proc. SPIE*, **2089**, 32–37, 1994.

Delbouille, L., Roland, G., and Gebbie, H. A., Actual possibilities of astronomical infrared spectroscopy by interferometry and Fourier transformation, *Ap. J.*, **69**, 334–338, 1964.

Delbouille, L., Roland, G., Brault, J. W., and Testerman, L., *Photometric Atlas of the Solar Spectrum from 1,850 to 10,000* cm^{-1}, Kitt Peak National Observatory Publication, National Optical Astronomy Observatory, 1983.

Delouis H., Fourier transformation of a 10^6 sample interferogram, *Aspen International Conference on Fourier Spectroscopy*, G. A. Vanasse, A. T. Stair, and D. J. Baker eds., U.S. Air Force Cambridge Research Laboratory, 71-0019, pp. 145–150, 1971.

de Zoeten, P., Haurer, R., and Birkl, R., Optical design of the Michelson Interferometer for Passive Atmospheric Sounding; Passive Infrared Remote Sensing of Clouds and the Atmosphere, *Proc. SPIE*, **1934**, 284, 1993.

Dolph, C. L., A Current Distribution for Broadside Arrays Which Optimizes the Relationship between Beam Width and Side-Lobe Level, *Proc. I.R.E.*, **34**, 335–348, 1946.

Dossier, B. *Rev. Opt.*, **33**, 57, 1954. The first suggestion to change an instrumental profile by suitable screening of a grating (apodization before the word was invented).

Elste, G., Die Entzerrung von Spektrallinien unter Verwendung von Voigtfunktionen, *Z. für Astrophysik*, **33**, 39–73, 1953.

Endemann, M., Lange, G., and Fladt, B., Michelson Interferometer for Passive Atmospheric Sounding; Space Optics 1994: Earth Observation and Astronomy, *Proc. SPIE*, **2209**, 36, 1994.

Fabry, C., and Perot, A, Théorie et application d'une nouvelle méthode de spectroscopie interférentielle, *Annales de Chimie et de Physique*, **16**, 115–44, 1899.

Faires, L. M., Fourier transforms for analytical atomic spectroscopy, *Anal. Chem.*, **58**, 1023A, 1986.

Farmer, C. B., *Can. J. Chem*, **52**, 1544–1559, 1974.

Farmer, C. B., High-resolution infrared spectroscopy of the Sun and the Earth's atmosphere from space, *Mikrochim. Acta [Wien]*, **III**, 189–214, 1987. [ATMOS]

Farmer, C. B., The ATMOS solar atlas, *Infrared Solar Physics*, 511–521, 1994.

Farmer, C. B., and Norton, R. H., Atlas of the infrared spectrum of the Sun and the Earth atmosphere from space. Volume I, The Sun, NASA Reference Publication 1224, 1989a.

Farmer, C. B., and Norton, R. H., Atlas of the infrared spectrum of the Sun and the Earth atmosphere from space. Volume II, Stratosphere and Mesosphere, 650 to 3350 cm^{-1}, NASA Reference Publication 1224, 1989b.

Farmer, C. B., Raper, O. F., and O'Callaghan, F. G., Final report on the first flight of the ATMOS instrument during the Spacelab 3 mission, April 29 through May 6, 1985, 1987, Jet Propulsion Laboratory Publication 87 32, 1987.

Farmer, C. B., Delbouille, L., Roland, G., and Servais, C., *The Solar Spectrum Between 16 and 40 microns*, San Juan Capistrano Research Inst. Publication; SJI Tech. Report 94-2, 1994.

Fellgett, P. B., *The Multiplex Advantage*, Ph.D. Thesis, University of Cambridge, Cambridge, 1951.

Fellgett, P. B., *J. Opt. Soc. Amer.*, **42**, 872, 1952a.

Fellgett, P. B., Symposium on Molecular Structure and Spectroscopy, Ohio State University, Columbus, OH, 1952b.

Fellgett, P. B., Spectrométre interfèrentiel multiplex pour mesures infra-rouge sur les étoiles, *Journal de Physique et le Radium*, **19**, 237–240, 1958.

Fellgett, P., Interventions regarding a paper by Beer and Cayford, *J. Physique*, **28**, C2-38–39, 1967.

Fellgett, P. B., The origins and logic of multiplex, Fourier, and interferometric methods in spectrometry, *Aspen International Conference on Fourier Spectroscopy*, G. A. Vanesse, A. T. Stair, and D. J. Baker eds., U.S. Air Force Cambridge Research Laboratory 71-0019, pp. 139–142, 1971.

Filler, A. H., Apodization and interpolation in Fourier-transform spectroscopy, *J. Opt. Soc. Am.*, **54**, 762, 1964. A concise treatise on apodization in which the concept of a *Filler diagram* (mapping sidelobe suppression *vs.* distortion of the central lobe) was introduced.

Filler, A. H., Some mathematical manipulations of interferograms, *J. Physique*, 28, C2–14, 1967.

Forman, M. L., Fast Fourier transform technique and its application to Fourier spectroscopy, *J. Opt. Soc. Amer.*, **56** 978–979, 1966.

Forman, M. L., *Journal de Physique*, **28**, C2-58, 1967. Introduction of the FFT algorithm to the FTS community.

Forman, M. L., Steel, W. H., and Vanasse, G. A., Correction of asymmetric interferograms obtained in Fourier spectroscopy, *J. Opt. Soc. Amer.*, **56**, 59–63, 1966.

Gebbie, H. A., *Phys. Rev.*, **107**, 1194, 1957.

Gebbie, H. A., *J. Phys. Radium*, **19**, 230, 1958. The first suggestion of numerical apodization in the computation process.

Gebbie, H. A., and Vanasse, G., *Nature*, **178**, 432, 1956.

Gebbie, H. A., Vanasse G., and Strong, J., *J. Opt. Soc. Amer.*, **46**, 377, 1956.

Geller, M., Atlas of the infrared spectrum of the Sun and the Earth atmosphere from space. Volume III, Key to Identification of Solar Features, NASA Reference Publication 1224, 1989.

Genest, J., and Tremblay, P., Instrument line shape of Fourier transform spectrometers: analytic solutions for nonuniformly illuminated off-axis detectors, *Appl. Opt.*, **38**, 5438–5446, 1999.

Graham, J. R., Abrams, M. C., Bennett, C. L., Carr, J., Cook, K., Dey, A., Wishnow, E., The performance and scientific rational for an IR imaging Fourier transform spectrograph on a large space telescope, *Proc. Astro. Soc. Pac.*, **110**, 1205, 1998.

Griffin, M. J., Bolometer instrument for FIRST, *Proc. SPIE*, Advanced Technology, MMW, Radio, and Terrahertz Telescopes, T. G. Phillips ed., **3357**, 45, 1998.

Griffiths, P. R., and de Haseth, J. A., in *Fourier Transform Infrared Spectrometry*, P. J. Elving and J. D. Winefordner, eds., Wiley–Interscience, New York, 1986.

Guelachvili, G., High–accuracy Doppler–limited 10^6 samples Fourier transform spectroscopy, *Appl. Opt.*, **17**, 1322–1326, 1978. 0.001-cm^{-1} resolution spectra.

Guelachvili, G., Distortion-free interferograms in Fourier transform spectroscopy with nonlinear detectors, *Appl. Opt.*, **25**, 4644–4648, 1986. A method of reducing the nonlinear effects by using the two outputs of a FTS instrument (analogous to the virtues of a "push–pull" method to minimize nonlinearities in a power amplifier).

Guelachvili, G., and Maillard, J.-P., Fourier spectroscopy from 10^6 samples, *Aspen International Conference on Fourier Spectroscopy*, G. A. Vanasse, A. T. Stair, and D. J. Baker eds., U.S. Air Force Cambridge Research Laboratory, 71-0019, 151–161, 1971.

Guelachvili, G., Birk, M., Borde, C. J., Brault, J. W., Brown, L. R., Carli, B., Cole, A. R. H., Evenson, K. M., Fayt, A., Hausamann, D., Johns, J. W. C., Kauppinen, J., Kou, Q., Maki, A. G., Narahari Rao, K., Toth, R. A., Urban, W., Valentin, A., Verges, J., Wagner, G., Wappelhorst, M. H., Wells, J. S., Winnewisser, B. P., and Winnewisser, M., High-resolution wavenumber standards for the infrared, *Spectrochimica Acta, Part A (Molec. Biomolec. Spectrosc.)*, **52A** , 7, 717–32, 1996.

Gush, H. P., and Buijs, H. L., *Can. J. Phys.*, **42**, 1037, 1964. Balloon-borne FTS for observations of emission by OH in the night sky.

Hall, D. N. B., Ridgway, S. T., Bell, E. A. and Yarborough, J. M., A 1.4 Meter Fourier Transform Spectrometer for astonomical observations, *Proc. SPIE*, **172**,121–129, 1979.

Hanel, R. A., Recent advances in satellite radiation measurements, in *Advances in Geophysics*, H. E. Landsberg and J. Van Mieghem eds., Academic Press, New York, 1970.

Hanel, R. A., and Chaney, L., Goddard SFC, Rep.X-620-66-476, 1966. NIMBUS III IRIS, with a spectral resolution of 5 cm^{-1}, covering 400 to 2000 cm^{-1}.

Hanel, R. A., and Conrath, B., Interferometer experiment on Nimbus 3: preliminary results, *Science*, **204**, 972–976, 1969.

Hanel, R. A., Schlachman, B., Clark, F. D., Prokesh, C. H., Taylor, J. B., Wilson, W. M., and Chaney, L., The Nimbus 3 Michelson interferometer, *Appl. Opt.*, **9**, 1767–1774, 1970.

Hanel, R. A., Schlachman, B., Rodgers, D., and Vanous, D., The Nimbus 4 Michelson interferometer, *Appl. Opt.*, **10**, 1376–1381, 1971.

Hanel, R. A., Conrath, B. J., Kunde, V. G., Prabhakara, C., Revah, I., Salomonson, V. V., and Wolford, G., The Nimbus 4 infrared spectroscopy experiment, I. Calibrated thermal emission spectra, *J. Geophys. Res.*, **77**, 2639–2641, 1972a.

Hanel, R. A., Schlachman, B., Breihan, E., Bywaters, R., Chapman, F., Rhodes, M., Rogers, D., and Vanous, D., The Mariner 9 Michelson interferometer, *Appl. Opt.*, **11**, 2625–2634, 1972b.

Hanel, R. A., Conrath, B. J., Hovis, W. A., Kunde, V. G., Lowman, P. D., Pearl, J. C., Prabhakara, C., and Schlachman, B. L., Infrared spectroscopy experiment on the Mariner 9 mission: preliminary results, *Science*, **175**, 305–308, 1972c.

Hanel, R. A., Conrath, B. J., Flasar, M., Kunde, V. G., Lowman, P., Maguire, W., Pearl, J., Pirraglia, J., and Samuelson, R., Infrared observations of the Jovain system from Voyager I, *Science*, **204**, 972–976, 1979.

Hanel, R. A., Crosby, D., Herath, L., Vanous, D., Collins, D., Creswick, H., Harris, C., and Rhodes, M., Infrared spectrometer for Voyager, *Appl. Opt.*, **19**, 1391–1400, 1980.

Hanel, R. A., Conrath, B. J., Flasar, F. M., Kunde, V. G., Maguire, W., Pearl, J., Pirraglia, J., Samuelson, R., Cruikshank, D., Gautier, D., Gierasch, P., Horn, L., and Schulte, P., Infrared observations of the Uranian system, *Science*, **233**, 70–74, 1986.

Hanel, R. A., Conrath, B. J., Jennings, D. E., and Samuelson, R. E., *Exploration of the Solar System by Infrared Remote Sensing*, Cambridge University Press, Cambridge, England, 1992. The grand tour of infrared remote sensing by members of the team that sent FTS instruments to Earth, Mars, Jupiter, Saturn, Uranus, and Neptune and numerous satellites in the process.

Harris, F. J., On the use of windows for harmonic analysis with the discrete Fourier transform, *Proc. IEEE*, **66**, 51–83, 1978. A vast expansion of the work done by Filler with the introduction of a signal-to-noise gain metric. Apodization windows include: rectangle, triangle, Hanning ($\cos^2$), Hamming, Rietz, Riemann, de la Valle–Poussin, Tukey, Bohman, Poisson, Hanning–Poisson, Cauchy, Gaussian, Dolph–Chebyshev, Kaiser–Bessel, Barcilon–Temes, exact Blackman, and Blackman–Harris.

Harrison, G. R., and Molnar, J. P., *J. Opt. Soc. Amer.*, **30**, 343, 1940.

Helms, H. D., Digital filters with equiripple or minimax responses, *IEEE Trans Audio Electroacoust.*, **AU-19**, 87–94, 1971.

Husson, N., Bonnet, B., Scott, N. A., and Chedin, A., Management and study of spectroscopic information: the GEISA program, *J. Quant. Rad. Spectrosc. Radiat Transfer*, **48**, 509–518, 1992.

Irfan, A. Y., Thorne, A. P., Bohlander, R. A., Gebbie, H. A., Fourier transform spectroscopy in the vacuum ultraviolet, *J. Physics E (Scientific Instruments)*, **12**, 6 472–473, 1979.

Jacquinot, P., Quelques recherches sur les raise faibles dans les spectres optiques, *Proc. Phys. Soc.*, **63**, 12, 969–979, 1950.

Jacquinot, P., *The Etendue Advantage*, XVII me Congrs du GAMS, Paris, 1954.

Jacquinot, P, The luminosity of spectrometers with prisms, gratings, or Fabry-Perot etalons, *J. Opt. Soc. Amer.* 44, 761, 1954.

Jacquinot, P., *J. Phys. Radium*, **19**, 223, 1958.

Jacquinot, P., New developments in interference spectroscopy, *Rep. Progress Physics*, **XXIII**, 295–306, 1960. The proccedings of a meeting held in 1957, in Bellevue, France, that provide a complete and excellent review of the subject, summarizing the status of various developments at the time and giving references.

Jacquinot, P., and Dufour, C., Condition optique d'emploi des cellules photo-électriques dans les spectrographes et les interfèrométres, *Journal Recherche du Centre National Recherche Scientifique Laboratorie Bellevue (Paris)*, **6**, 91–103, 1948.

Javelle, P., and Cayla, F., Infrared Atmospheric Sounding Interferometer (IASI) instrument overview; Space Optics 1994: Earth Observation and Astronomy, *Proc. SPIE*, **2209**, 14, 1994.

Jennings, D. E., Hubbard, R., and Brault, J. W., Double passing the Kitt Peak 1-Meter Fourier transform spectrometer, *Appl. Opt.*, **24**, 3438–3440 (1985).

Johns, J. W., High resolution and the accurate measurement of intensities, *Mikrochim. Acta [Wien]*, **III**, 171–188, 1987.

Kawada, M., FIS: Far-Infrared Surveyor onboard IRIS, *Proc. SPIE*, Infrared Astronomical Instrumentation, A. M. Fowler ed., **3354**, 6, 1998.

Kunde, V. G., Brasunas, J. C., Conrath, B. J., Hanel, R. A., Herman, J. R., Jennings, D. E., Maguire, W. C., Walser, D. W., Annen, J. N., Silverstein, M. J., Abbas, M. M., Herath, L. W., Buijs, H. L., Berube, J. N., and McKinnon, J., Infrared spectroscopy of the lower atmosphere with a balloon-borne Fourier spectrometer, *Appl. Opt.*, **26**, 545–553, 1987. Mid-infrared limb emission atmospheric sounding with a cryogenic FTS.

Kunde, V. G., *et al.* Cassini infrared Fourier spectroscopic investigation, *Proc. SPIE*, **2803**, Cassini/Huygens: A Mission to the Saturnian Systems, Linda Horn; ed., pp. 162–177, 1996.

Learner, R. C. M. and Thorne, A. P., Wavelength calibration of Fourier transform emission spectra with applications to Fe I, *J. Opt. Soc. Amer. B* , **5**, 2045–2049, 1988.

Learner, R. C. M., Thorne, A. P., Wynne-Jones, I., Brault, J. W., and Abrams, M. C., Phase correction of emission line Fourier transform spectra, *J. Opt. Soc. Amer. A*, **12**, 10, 2165–2171, 1995.

Learner, R. C. M., Thorne, A. P., Brault, J. W., Ghosts and artifacts in Fourier-transform spectrometry, *Appl. Opt.*, **35**, 2947–2954, 1996.

Lee, J. P. and Comisarow, M. B., Apodization Functions for Absorption-Mode Fourier Transform Spectroscopy, *Appl. Spec.*, **43**, 599–604, 1989.

Lee, Y. W., *Statistical Theory of Communications*, Wiley, New York, 1960.

Maillard, J.-P., 3D Spectroscopy with a Fourier transform spectrometer, in *3D Optical Spectroscopic Methods in Astronomy*, ASP Conference Series, G. Comte and M. Marcelin eds., **71**, 316, 1995.

Maillard, J.-P., and Michel, G., A high-resolution Fourier transform spectrometer for the cassegrain focus at the CFH telescope, in *Instrumentation for Astronomy with Large Optical Telescopes*, C. M. Humphries ed., pp. 213–222, Kluwer Academic/Reidel, Norwell, MA, 1982.

Mather, J., and Kelsall, T., The Cosmic Background Explorer Satellite, *Physica Scripta*, **21**, 671–677, 1980.

Mather, J. C., Fixsen D. J., and Shafer, R. A., Design for the COBE Far Infrared Absolute Spectrophotometer (FIRAS), *Proc. SPIE*, **2019**, 168, 1993.

Mather, J. C., Cheng, E. S, Cottingham, D. A., Eplee, R. E. Jr., Fixsen, D. J., Hewagama, T., Isaacman, R. B., Jensen, K. A., Meyer, S. S., Noerdlinger, P. D., Read, S. M., Rosen, L. P., Shafer, R. A., Wright, E. L., Bennett, C. L., Boggess, N. W., Hauser, M. G., Kelsall, T., Moseley, S. H. Jr., Silverberg, R. F., Smoot, G. F., Weiss, R., and Wilkins, D. T., Measurement of the cosmic microwave background spectrum by the COBE FIRAS instrument, *Astrophys. J.*, **420**, 439, 1994.

Maymon, P. W., Dittman, M. G., Pasquale, B. A., Jennings, D. E., Mehalick, K. I., Trout, C. J., Optical design of the Composite Infrared Spectrometer (CIRS) for the Cassini mission, *Proc. SPIE*, **1945**, Space Astronomical Telescopes and Instruments II, P. Y. Bely, and J. B. Breckinridge eds. pp. 100–111, 1993.

Mertz, L. W., *J. Phys. Radium*, **19**, 233, 1958. A polarization interferometer for the visible spectral region consisting of a double refracting blade of variable thickness between the polarizers.

Mertz, L. W., in *Infrared Spectra of Astronomical Bodies* (Proc. 12th Intern. Astrophys. Symp., Liège, Belgium, June 1963 (Institute d'Astrophysique, Cointe-Sclessin, Belgium, 1964), p. 120, 1963. Rapid scanning.

Mertz, L., *Transformations in Optics,* Wiley, New York, 1965.

Mertz, L. W., *J. Phys.*, **28**, C2-11, 1967a. Initial discussion of phase error.

Mertz, L., Auxiliary computation for Fourier spectroscopy, *Infrared Phys.*, **7**, 17, 1967b.

Mertz, L. W., *Infrared Physics*, **7**, 17, 1967c. Phase correction.

Mertz, L. W., Fourier spectroscopy, past, present, and future, *Appl. Opt.*, **10**, 386–389, 1971.

Michelson, A. A., The relative motion of the Earth and the luminiferous ether, *Am. J. Sci.*, 3, **22**, 120–129, 1881.

Michelson, A. A., *Studies in Optics*, Chicago, University of Chicago Press, Chicago, IL, 1927. In particular see the chapter entitled "Visibility of Fringes."

Michelson, A. A., and Morley, E. W., On the relative motion of the Earth and the luminiferous ether, *Am. J. Sci.*, 3, **34**, 333- 45, 1887.

Michelson, A. A., and Stratton, S. W., A new harmonic analyzer, *Am. J. Sci.*, 4, **5**, 1–13, 1898.

Mill, J. D., Midcourse Space Experiment (MSX), an overview of instruments and data collection plan: signal processing, sensor fusion, and target recognition III, *Proc. SPIE*, **2232**, 200, 1994.

Nave, G., Learner, R. C. M., Thorne, A. P., and Harris, C. J., Precision Fe I and Fe II wavelengths in the ultraviolet spectrum of the iron-neon hollow cathode lamp, *J. Opt. Soc. Amer. B*, **8**, 2028–2041, 1991.

Nave, G., Learner, R. C. M., Murray, J. E., Thorne, A. P., and Brault, J. W., Precision Fe I and Fe II wavelengths in the red and infra-red spectrum of the iron-neon hollow cathode lamp, *J. Phys. II (France)*, **2**, 913–929, 1992.

Norton, R. H., and Beer, R., New apodizing functions for Fourier spectrometry, *J. Opt. Soc. Amer.*, **66**, 259–264, 1976; **67**, 419, 1977. The erratum in volume **67** is rarely referenced.

Nyquist, H., Certain topics in telegraph transmission theory, *A.I.E.E. Trans.*, **617**, 644, 1928.

Persky, M. J., A review of spaceborne infrared Fourier transform spectrometers for remote sensing, *Rev. Sci. Instrum.*, **66**, 10, 4763–4797,1995.

Pinard, J., *J. Phys.*, **28**, C2-136, 1967. The first very high-resolution spectra (0.006 cm^{-1}).

Rabiner, J., ed., *Programs for Digital Signal Processing*, IEEE Press, New York, 1972.

Rayleigh, Lord, *Phil. Mag.*, **8**, 261, 1879.

Ridgway, S.T. and Brault, J. W., Astronomical Fourier transform spectroscopy revisited, *Ann. Rev. Astron. Astrophys.*, **22**, 291–317, 1984. A review that includes references to other reviews as well as many original papers.

Rothman, L. S., Goldman, A., Gillis, J. R., Gamache, R. R., Pickett, H. M., Poynter, R. L., Husson, N., and Chedin, A., AFGL trace gas compilation — 1982 version, *Appl. Opt.*, **22**, 1616– 1627, 1983a.

Rothman, L. S., Gamache, R. R., Barbe, A., Goldman, A., Gillis, J. R., Brown, L. R., Toth, R. A., Flaud, J.-M., and Camy-Peyret, C., AFGL atmospheric absorption-line parameters compilation — 1982 edition, *Appl. Opt.*, **22**, 2247–2256, 1983b.

Rothman, L. S., Gamache, R. R., Goldman, A., Brown, L. R., Toth, R. A., Pickett, H. M., Poynter, R. L., Flaud, J.-M., Camy-Peyret, C., Barbe, A., Husson, N., and Rinsland, C. P., and Smith, M. A. H., The HITRAN database: 1986 edition, *Appl. Opt.*, **26**, 4058–4097, 1987.

Rothman, L. S., Gamache, R. R., Tipping, R. H., Rinsland, C. P., Smith, M. A. H., Benner, D. C., Malathy Devi, V., Flaud, J.-M., Camy-Peyret, C., Perrin, A., Goldman, A., Massie, S. T., Brown, L. R., and Toth, R. A., The HITRAN molecular database editions of 1991 and 1992, *Appl. Opt.*, **26**, 4058–4097, 1992.

Rothman, L. S., Rinsland, C. P., Goldman, A., Massie, S. T., Flaud, J.-M., Perrin, A., Dana, V., Mandin, J.-Y., Schroeder, J., Mc Cann, A., Gamache, R. R., Wattson, R. B., Yoshino, K., Chance, K., Jucks, K., Brown, L. R. and Varanasi, P., The

HITRAN molecular spectroscopic database and HAWKS (HITRAN Atmospheric Workstation), *J. Quant. Spectrosc. Radiat. Transfer*, **60**, 665–682, 1998.

Rubens, H., and Wood, R. W., *Phil. Mag.*, **21**, 249, 1911

Schinder, R. A., *Appl. Opt.*, **9**, 301–306, 1970.

Savitsky, A., and Golay, M. J. E., Smoothing and differentiation of data by simplified least squares procedures, *Anal. Chem.*, **36**, 1627, 1964. One of the few historical references on *line finding* by differentiation. The procedures and conclusions are not numerical optimal.

Shimoda, H., and Ogawa, T., Interferometric Monitor for Greenhouse Gases (IMG); Infrared Spaceborne Remote Sensing II, *Proc. SPIE*, **2268**, 92, 1994.

Steel, W. H., *Interferometry*, Cambridge University Press, Cambridge, MA, 1983. An overview of interferometry (not just for FTS) by one of the pioneers.

Strong, J., *J. Opt. Soc. Amer.*, **47**, 354, 1957.

Thorne, A. P., Harris, C. J., Wynne-Jones, I., Learner, R. C. M., Cox, G., A Fourier transform spectrometer for the vacuum ultraviolet: design and performance, *J. Physics E (Scientific Instruments)*, **20**, 54–60, 1987.

Thorne, A. P., Fourier transform spectrometry in the ultraviolet, *Anal. Chem.*, **63**, 57A, 1991.

Thorne, A. P., *Spectrophysics*, 2nd ed., Chapman and Hall, New York, 1988.

Thorne, A. P., Litzen, U., and Johansson, S., *Spectrophysics: Principles and Applications*, Springer-Verlag, New York, 1999.

Toth, R. A., Farmer, C. B., Schindler, R. A. Raper, O. F., Shaper, P. W., *Nature (London) Phys. Sci.*, **244**, 7–8, 1973.

Toon, G. C., The JPL MkIV interferometer, *Optics and Photonics News*, **2**, 19-21, October 1991.

Traub, W. A., Chance, K. V., Brasunas, J. C., Vrtilek, J. M., and Carleton, N. P., Use of a Fourier transform spectrometer on a balloon-borne telescope and at the Multiple Mirror Telescope (MMT), *Proc. SPIE*, **331**, 208–218, 1982.

Van de Hulst, H. C., and Reesinck, J. J., *Astrophysic. J.*, **106**, 121, 1947.

Vaughan, M., *The Fabry–Perot Interferometer*, Institute of Physics, New York, 1989.

Whaling, W., Anderson, W. H. C., Carle, M. T., Brault, J. W., Zarem, H. A., Argon ion linelist and level energies in the hollow-cathode discharge, *J. Quant. Spectrosc. Rad. Trans.*, **53**, 1–22, 1995.

Wiener, N., *The Extrapolation, Interpolation, and Smoothing of Statistical Time Series*, Wiley, New York, 1949.

Wood, R. W., *Physical Optics*, McMillan, New York, 3rd ed., 1934.

Woody, D. P., Mather, J. C., Nishioka, N. S., and Richards, P. L., Measurement of the spectrum of the submillimeter cosmic background, *Phys. Rev. Lett.*, **34**, 1036–1039, 1975.

Zachor, A., *et al.*, Mid-course Space Experiment (MSX): Capabilities of the LWIR interferometer for remote sensing of trace constituents in the stratosphere and mesosphere; Atmospheric Propagation and Remote Sensing, *Proc. SPIE*, **2222**, 99, 1994.

Ziessov, D., *On-line Rechner in der Chemie. Grundlagen und Anwendung in der Fourier Spektroskopie*, Walter de Gruyter, Berlin, pp. 345–354, 1973. An considerable expansion of the work of Savitisky and Golay with numerous corrections.

Index

Printed and bound by CPI Group (UK) Ltd, Croydon, CR0 4YY
08/05/2025
01864786-0001